T0201536

Multicast in Third-Generation Mobile Networks

Multicast in Third-Generation Mobile Networks

Services, Mechanisms and Performance

Robert Rümmler
Accenture AG, Switzerland

Alexander Gluhak
LM Ericsson Limited, Republic of Ireland

A. Hamid Aghvami
King's College London, United Kingdom

A John Wiley and Sons, Ltd., Publication

Library of Congress Cataloging-in-Publication Data

Rummler, Robert.
   Multicast in third-generation mobile networks : services, mechanisms, and preformance /
Robert Rummler, Alexander Daniel Gluhak, Hamid Aghvami.
      p. cm.
   Includes bibliographical references and index.
   ISBN 978-0-470-72326-5 (cloth)
   1. Multicasting (Computer networks) I. Gluhak, Alexander Daniel. II. Aghvami, Hamid.
III. Title.
   TK5105.887.R86 2009
   004.6'6–dc22

                                                                                    2008049882

British Library Cataloguing in Publication Data

A catalogue record for this book is available from the British Library

ISBN 978-0-470-72326-5 (H/B)

Typeset in 10/12 Times by Laserwords Private Limited, Chennai, India
Printed and bound in Great Britain by CPI Antony Rowe, Chippenham, Wiltshire

Contents

Biography

Robert Rümmler

Robert Rümmler received his Bachelor and PhD degrees in electrical engineering from King's College, University of London, in 2000 and 2005 respectively. From 2001 to 2005 he worked as a research associate at the Centre for Telecommunications Research, King's College London, and contributed to several European research projects in the area of reconfigurability. He currently works as a consultant for Accenture AG, Zürich, Switzerland. His research interests are in multicast for third-generation networks, software-defined radio and end-to-end reconfigurability. He has authored many papers in refereed international conferences and journals. He is a member of the IEEE.

Alexander Gluhak

Alexander is currently a researcher at the Ericsson Ireland Research Centre. He received a Dipl.-Ing. (FH) in communications engineering from the University of Applied Sciences Offenburg, Germany, in 2002 and a PhD in mobile networks and systems from the University of Surrey, UK, in 2006. His research interests are in mobile multicast delivery, heterogeneous network architectures, and scalable context information infrastructures for next generation networks. He has won several awards for his research contributions, such as the Deutsche Telekom Award in 2002, the JSPS Research Fellowship Award in 2005 and the MobileVCE Research Award 2006.

A. Hamid Aghvami

Hamid Aghvami obtained his MSc and PhD degrees from King's College, The University of London, in 1978 and 1981 respectively. He joined the academic staff at King's in 1984. In 1989 he was promoted to Reader and in 1993 was promoted Professor in Telecommunications Engineering. He is presently the Director of the Centre for Telecommunications Research at King's. Professor Aghvami carries out consulting on digital radio communication systems for British and international companies. He has published over 480 technical papers and given invited talks all over the world on various aspects for personal and mobile radio communications as well as giving courses on the subject world wide. He was a Visiting Professor at NTT Radio Communication Systems Laboratories in 1990 and and a Senior Research Fellow at BT Laboratories in 1998–1999. He was an Executive Advisor to Wireless Facilities Inc., USA in 1996–2002. He is the Managing Director of Wireless Multimedia Communications Ltd, his own consultancy.

He leads an active research team working on numerous mobile and personal communications projects for future-generation systems, these projects are supported both by the government and industry. He was a member of the Board of Governors of the IEEE Communications Society in 2001–2003. He is a distinguished lecturer of the IEEE Communications Society and has been member, Chairman, and Vice-Chairman of the technical programme and organizing committees of a large number of international conferences. He is also founder of the international conference on Personal Indoor and Mobile Radio Communications (PIMRC). He is a Fellow of the Royal Academy of Engineering, Fellow of the IET and Fellow of the IEEE.

Preface

This book investigates the deployment of multicast in third-generation mobile networks. Multicast is the delivery of data to a group of destinations simultaneously, using the most efficient strategy to deliver the data. The book gives an overview of the services that can be realized with multicast in third-generation networks, describes the mechanisms required to support these services and evaluates the performance of several mechanisms for multicast. The focus of this book is on multicast in Universal Mobile Telecommunication System (UMTS) and CDMA2000 networks, the dominant third-generation network standards. The authors hope to provide a good balance between describing the relevant mechanisms for multicast in third-generation networks, providing useful considerations and presenting specific research results.

The book is structured as follows. Chapters 1 to 3 provide an overview of cellular mobile communication systems, the fundamentals of multicast in IP networks and the most relevant aspects of third-generation mobile networks. Chapter 4 discusses some of the services that may be realized with multicast in third-generation networks. Chapters 5 to 7 explore the multicast capabilities of third-generation networks. The Multimedia Broadcast/Multicast Service (MBMS) standard for multicast in UMTS and the Broadcast/Multicast Service (BCMCS) standard for multicast in CDMA2000 networks are described in detail. Chapters 8 to 10 cover the performance of multicast in third-generation networks in terms of radio capacity, multicast routing cost to the network, as well as the efficiency of reliable multicast with respect to throughput and delay. Chapter 11 finally presents mechanisms for delivering multicast in a heterogeneous network environment in which third-generation mobile technology coexists with digital broadcast technology.

Chapter 1 introduces the main concepts of mobile cellular communication systems, describes some important fundamentals of data networking and briefly outlines how multicast can be achieved in data networks as well as in cellular mobile networks.

Chapter 2 introduces the fundamentals of IP multicast. The chapter first provides an overview of the IP multicast service model as well as multicast addressing, followed by a review of the mechanisms for multicast address assignment and session announcement. Group management and routing for IP multicast are then described in detail. As a more advanced topic, protocols and mechanisms for reliable multicast delivery are detailed. Also, congestion and flow control for IP multicast are briefly touched upon, followed by a brief introduction of solutions that support multicast in a mobile environment.

Chapter 3 describes the most important aspects of UMTS and CDMA2000 third-generation networks. The chapter describes the air interface, the radio and core networks of UMTS and CDMA2000, as well as several relevant procedures such as mobility and session management.

Chapter 4 provides an overview of mobile services that can be realized with multicast in third-generation networks. Several services that can be offered with multicast are described in the form of use cases. Several high-level requirements that the system must provide in order fully to support the described services are extracted from the use cases. In addition, the factors that have an influence on whether multicast services will be accepted by users and succeed in the marketplace are discussed.

Chapter 5 explores the multicast capabilities of third-generation networks. For both UMTS and CDMA2000 networks, the network extensions for MBMS and BCMCS that support the efficient delivery of multicast traffic are introduced. The modifications to the radio access and core network architecture for MBMS and BCMCS are outlined. The chapter also provides an overview of the different multicast service delivery phases within UMTS and CDMA2000 networks.

Chapter 6 describes the MBMS standard in detail. It explains the different procedures that are relevant for the management of MBMS bearer services and discusses practical issues in routing multicast packets on the bearer path. Additionally, the MBMS service layer and its mechanisms are described.

Chapter 7 focuses on the BCMCS standard. The chapter covers the BCMCS network layer as well as the bearer service architecture and its management. The chapter also describes the service layer of BCMCS in detail.

Chapter 8 analyses the capacity for performing multicast over the CDMA air interface. In CDMA, data transfer to a group may either take place on multiple point-to-point channels transmitted to individual multicast users separately or on a single point-to-multipoint channel that is broadcast over the entire cell. The chapter provides insight into the trade-offs between employing point-to-point and point-to-multipoint channels for multicast over the CDMA air interface.

Chapter 9 investigates the cost of packet delivery and location update cost of different mechanisms for multipoint data transfer in UMTS networks. Firstly, an alternative mechanism for performing multicast routing in UMTS networks is described. Then, cost expressions for the packet delivery and location update cost of several mechanisms for multipoint data transfer such as MBMS are derived. Finally, the performance trade-off between the proposed mechanism for routing multicast packets in UMTS is evaluated numerically and compared against that of MBMS and other viable mechanisms for multicast data transfer.

Chapter 10 investigates the performance of different reliability mechanisms for multicast. Reliability mechanisms that combine packet-based forward error correction with automatic repeat request are considered. The performance of these mechanisms, applied both to the radio link control layer as well as the application layer, are evaluated in terms of channel efficiency, residual packet error rate and delay.

Chapter 11 takes a look at alternative technologies for mobile multicast delivery. Several wireless communication technologies such as DVB-H, MediaFlo, ISDB-T and T-DMB that are suitable for multicast service delivery are reviewed. The motivation and benefits of using these technologies for multicast service delivery are described. Also, the challenges in providing multicast services in a heterogeneous network environment consisting of different network technologies are discussed. The chapter introduces a potential approach for achieving coordinated delivery of multicast services in a heterogeneous environment consisting of several wireless networks. Chapter 11 concludes the main body of this book.

Chapter 11 is followed by two appendices. Appendix A derives the closed-form expressions for the capacity of employing dedicated point-to-point and shared point-to-multipoint channels for multicast over the CDMA air interface. The numerical evaluation of these expressions is presented in Chapter 8. Appendix B derives the cost expressions for the packet delivery and location update cost of different mechanisms for performing multipoint data transfers in UMTS. These cost expressions are evaluated numerically in Chapter 6.

Each chapter is preceded by a short outline of the topics to be treated and closes with a summary and some intermediate conclusions. Readers may find the list of abbreviations and a list containing the mathematical symbols used throughout the book useful. The list of abbreviations and the list of symbols can be found after the acknowledgements. As far as acronyms are concerned, an effort has been made to write them out in full whenever they occur first in each chapter. Exceptions to this rule include regularly recurring acronyms and cases where acronyms are used in passing first and explicitly introduced soon after.

Acknowledgements

The authors would like to thank the many people who provided support or directly contributed to the research efforts documented in this book. In particular, the research contributions of Yun Won Chung and Imran Ashraf are greatly appreciated.

Robert Rümmler would like to thank his parents for all the support provided. Words cannot express enough gratitude. Many thanks also to his brother Richard and his uncle Wolfgang. Paul Pangalos was instrumental in the writing of this book by suggesting Alex Gluhak as a coauthor. Ben Allen and Maciej Nawrocki were very helpful in answering a number of questions in the early stages of the book. For encouragement and moral support, thanks to Katharina, Uwe, Kieran and Thorsten.

Alex Gluhak would like to thank his family for their continuous support throughout his life. He would like to thank his wife Monica, in particular, for her understanding and many sacrifices, considering the countless weekends that were spent writing the book. Special thanks go to Paul Pangalos for establishing the contact for the coauthorship of this book, and to JP, whose company provided a refreshing change on some of the dull and rainy weekends.

The team at John Wiley & Sons participating in the production of this book provided excellent support. The authors would like to thank Sarah Hinton for her assistance with many practical issues in the production process, Anna Smart for the cover design and especially the copy editor for his efforts in correcting errors and improving the overall quality of the book.

List of Abbreviations

3GPP	Third-Generation Partnership Project
3GPP2	Third-Generation Partnership Project 2

A

AAA	Authentication, Authorization and Accounting
AAC	Advanced Audio Coding
AC	Asynchronous Control
ACK	Acknowledgement
AI	Acquisition Indicator
AICH	Acquisition Indicator Channel
ALC	Asynchronous Layered Coding
AMPS	American Mobile Phone System
AN	Access Network
AP-AICH	Access Preamble AICH
API	Access Preamble Indicator
APN	Access Point Name
ARP	Address Resolution Protocol
ARPU	Average Revenue Per User
ARQ	Automatic Repeat Request
ASIC	Application-Specific Integrated Circuit
ASM	Any Source Multicast
AT	Access Terminal
ATM	Asychronous Transfer Mode

B

BAK	Broadcast Access Key
BCCH	Broadcast Control Channel
BCH	Broadcast Channel
BC/MC	Broadcast/Multicast
BCMCS	Broadcast/Multicast Service
BCMCS-C	BCMCS Controller
BGMP	Border Gateway Multicast Protocol
BGP	Border Gateway Protocol
BMC	Broadcast/Multicast Control
BM-SC	Broadcast/Multicast Service Centre

BPSK	Binary-Phase Shift-Keying
BS	Base Station
BSC	Base Station Controller
BSF	Bootstrapping Service Function
BSN	Broadcast Serving Node
BTS	Base Station Transceiver System

C

CBC	Cell Broadcast Centre
CBE	Cell Broadcast Entity
CBS	Cell Broadcast Service
CBT	Core-Based Trees
CCC	Content Casting Centre
CCCH	Common Control Channel
CCTrCH	Coded Composite Transport Channel
CD/CA-ICH	Collision Detection/Channel Assignment Indicator Channel
CDI	Collision Detection Indicators
CDI/CAI	Collision Detection/Collision Assignment Indicators
CDMA	Code-Division Multiple Access
CN	Core Network
CoA	Care of Address
CPCH	Common Packet Channel
CPHCH	Common Physical Channel
CPICH	Common Pilot Channel
CRC	Cyclic Redundancy Check
CS	Content Server
CSI	Channel State Information
CSICH	CPCH Status Indicator Channel
CTCH	Common Traffic Channel

D

DA	Destination Address
DAB	Digital Audio Broadcasting
D-AMPS	Digital AMPS
DB	Dynamic Broadcast
DCCH	Dedicated Control Channel
DCH	Dedicated Channel
DECT	Digital Enhanced Cordless Telecommunications
DMB	Digital Multimedia Broadcasting
DMSP	Designated Multicast Service Provider
DPCCH	Dedicated Physical Control Channel
DPCH	Dedicated Physical Channel
DPDCH	Dedicated Physical Data Channel
DPHCH	Dedicated Physical Channel
DRC	Data Rate Control
DRNC	Drift RNC
DS	Direct-Sequence
DSCH	Downlink Shared Channel
DSP	Digital Signal Processing

DTCH	Dedicated Traffic Channel
DTMC	Discrete-Time Markov Chain
DVB	Digital Video Broadcasting
DVB-H	Digital Video Broadcasting Handheld
DVMRP	Distance-Vector Multicast Routing Protocol

E

EDGE	Enhanced Data Rates for Global Evolution
EPG	Electronic Programme Guide
ESG	Electronic Service Guide
ETSI	European Telecommunication Standards Institue
EV-DO	Evolution Data Only
EV-DV	Evolution Data Voice

F

FACH	Forward Access Channel
FDD	Frequency-Division Duplex
FDMA	Frequency-Division Multiple Access
FDT	File Delivery Table
FEC	Forward Error Correction
FH	Frequency-Hopping
FLUTE	File Delivery over Unidirectional Transport
FTP	File Transfer Protocol

G

GA	Group Address
GGSN	Gateway GPRS Support Node
GMMF	Group Membership Management Function
GPRS	General Packet Radio Service
GRE	Generic Routing Encapsulation
GSM	Global System for Mobile Communications
GTP	GPRS Tunnelling Protocol

H

HA	Home Agent
H-ARQ	Hybrid ARQ
HDLC	High-Level Data Link Control
HLR	Home Location Register
HRS	Home RADIUS Server
HSS	Home Subscriber Server
HTTP	Hypertext Transfer Protocol

I

IANA	Internet Assigned Numbers Authority
ID	Identifier
IEEE	Institute of Electrical and Electronics Engineers
IETF	Internet Engineering Task Force
IGMP	Internet Group Management Protocol
IGW	Interworking Gateway

iid	independent and identically distributed
IMSI	International Mobile Subscriber Identity
IMT-2000	International Mobile Telecommunications 2000
IP	Internet Protocol
IPE	IP Encapsulation
ISDB-T	Integrated Services Digital Broadcasting Terrestrial
ISDN	Integrated Services Digital Network
ITU	International Telecommunications Union

K

kbps	kilobits per second

L

LAC	Link Access Control
LAN	Local-Area Network
LCT	Layered Coding Transport
LTE	Long-Term Evolution

M

MA	Multicast Agent
MAAA	Multicast Address Allocation Architecture
MAC	Medium Access Control
MASC	Multicast Address Set Claim
MBGP	Multicast Border Gateway Protocol
MBMS	Multimedia Broadcast/Multicast Service
Mbone	multicast backbone
MCCH	MBMS Control Channel
ME	Mobile Equipment
MF	Multicast Forwarder
MFTP	Multicast File Transfer Protocol
MGA	Multicast Group Address
MGC	Multicast Group Context
MGV-S	MBMS Key Generation and Validation Storage
MH	Mobile Host
MHA	Multicast Home Agent
MICH	MBMS Notification Indicator Channel
MIKEY	Multimedia Internet Keying
MIME	Multipurpose Internet Mail Extensions
MLD	Multicast Listener Discovery
MM	Mobility Management
MMA	Multicast By Multicast Agent
MMS	Multimedia Messaging Service
MOSPF	Multicast OSPF
M-PDP	Multicast Packet Data Protocol
MPE	Multi-Protocol Encapsulation
MPEG	Moving Pictures Expert Group
M-RAB	Multicast RAB
MRK	MBMS Request Key
MS	Mobile Station

MSC	Mobile Switching Centre
MSCH	MBMS Scheduling Channel
MSDP	Multicast Source Discovery Protocol
MSISDN	Mobile Subscriber ISDN Number
MSK	Multicast Service Key
MSR	Multicast Subscriber Record
MT	Mobile Terminal
MTCH	MBMS Traffic Channel
M-TEID	Multicast Tunnel Endpoint Identifier
MTK	MBMS Traffic Key
MUK	MBMS User Key

N

NACK	Negative Acknowledgement
NMF	Network Management Function
NMT	Nordic Mobile Telephone
N-PDU	Network PDU
NSAPI	Network Layer Service Access Point Identifier

O

OFDM	Orthogonal Frequency-Division Multiplexing
OMA	Open Mobile Alliance
OSI	Open Systems Interconnection
OSPF	Open Shortest Path First
OVSF	Orthogonal Variable Spreading Factor

P

PCCH	Paging Control Channel
P-CCPCH	Primary Common Control Physical Channel
PCE	Power Control Error
PCF	Packet Control Function
PCH	Paging Channel
PCN	Packet Core Network
PCPCH	Physical Common Packet Channel
P-CPICH	Primary Common Pilot Channel
PDA	Personal Digital Assistant
PDC	Personal Digital Cellular
PDCP	Packet Data Convergence Protocol
PDN	Packet Data Network
PDP	Packet Data Protocol
PDSCH	Physical Downlink Shared Channel
PDSN	Packet Data Serving Node
PDU	Packet Data Unit
PGM	Pretty Good Multicast
PHS	Personal Handphone System
PI	Paging Indicator
PICH	Paging Indicator Channel
PID	Packet Identifier
PIM	Protocol-Independent Multicast

PIM-DM	Protocol-Independent Multicast Dense Mode
PIM-SM	Protocol-Independent Multicast Sparse Mode
PLMN	Public Land Mobile Network
PMM	Packet Mobility Management
PPP	Point-to-Point Protocol
PRACH	Physical Random Access Channel
PS	Packet-Switched
PSS	Packet-Switched Streaming
PSTN	Public Switched Telephone Network
PTM	Point-To-Multipoint
PTP	Point-to-Point

Q

QoE	Quality of Experience
QoS	Quality of Service
QPSK	Quadrature-Phase Shift-Keying

R

RA	Routing Area
RAB	Radio Access Bearer
RACH	Random Access Channel
RADIUS	Remote Authentication Dial-In User Service
RANAP	Radio Access Network Application Part
RBMoM	Range-Based Mobile Multicast
RF	Radio Frequency
RK	Registration Key
RLC	Radio Link Control
RLP	Radio Link Protocol
RM	Resource Manager
RMTP	Reliable Multicast Transport Protocol
RNC	Radio Network Controller
RNS	Radio Network Subsystem
RP	Rendezvous Point
RPB	Reverse-Path Broadcast
RPC	Reverse Power Control
RPM	Reverse-Path Multicasting
RRC	Radio Resource Control
RSS	Really Simple Syndication
RTCP	Real-Time Transport Control Protocol
RTP	Real-Time Protocol
RTSP	Real-Time Streaming Protocol
RTT	Radio Transmission Technology

S

SAP	Session Announcement Protocol
SB	Static Broadcast
SC	Sychronous Control
S-CCPCH	Secondary Common Control Physical Channel
SCF	Session Control Function

SCH	Sychronization Channel
S-CPICH	Secondary CPICH
SDP	Session Description Protocol
SDU	Service Data Unit
SF	Spreading Factor
SGSN	Serving GPRS Support Node
SI	Status Indicators
SIP	Session Initiation Protocol
SIR	Signal-to-Interference Ratio
SK	Short-Term Key
SM	Session Management
SMS	Short Message Service
SRBP	Signalling Radio Burst Protocol
SRM	Scalable Reliable Multicast
SRNC	Serving RNC
SRTP	Secure Real-Time Transport Protocol
SSM	Source-Specific Multicast
T	
TACS	Total Access Communication System
TAM	Technology Acceptance Model
TCP	Transmission Control Protocol
TDD	Time-Division Duplex
TDMA	Time-Division Multiple Access
T-DMB	Terrestrial Digital Multimedia Broadcast
TD-SCDMA	Time-Division Synchronous Code-Division Multiple Access
TEID	Tunnel Endpoint Identifier
TFC	Transport Format Combination
TFCI	Transport Format Combination Identifier
TG	Transmission Group
TI	Transaction Identifier
TK	Temporary Key
TMGI	Temporary Mobile Group Identity
TOI	Transmission Object Identifier
TRPB	Truncated Reverse-Path Broadcasting
TS	Transport Stream
TSI	Transport Session Identifier
TS-MUX	Transport Stream Multiplex
TTI	Transmission Time Interval
TTL	Time-To-Live
U	
UDP	User Datagram Protocol
UE	User Equipment
UHF	Ultrahigh-Frequency
UIM	User Identity Module
UMB	Ultra Mobile Broadband
UMTS	Universal Mobile Telecommunications System
URA	UTRAN Registration Area

USIM   User Subscriber Identity Module
UTRAN   UMTS Terrestrial Radio Access Network

V
VHF   Very High-Frequency

W
WAN   Wide-Area Network
WCDMA   Wideband CDMA
WLAN   Wireless Local-Area Network
WWW   World Wide Web

X
XCast   Explicit Multicast
XML   Extensible Mark-up Language

List of Symbols

A_{cell}	cell area
A_m	Erlang capacity for multicast
A_t	total network area
A_v	Erlang capacity for voice
C_{gs}	cost of MBMS registration/deregistration
C_{rr}	cost of UE linking procedure for MBMS
C_{sr}	cost of RNC registration/deregistration
D_{bu}	transmission cost of packet delivery over the air interface
D_{cell}	cost of packet delivery for cell-connected subscriber
D_{gs}	transmission cost of packet delivery between GGSN and SGSN
D_{rb}	transmission cost of packet delivery between RNC and Node B
D_{sr}	transmission cost of packet delivery between SGSN and RNC
D_I	packet delivery cost for broadcast
D_{II}	packet delivery cost of multiple unicast
D_{III}	packet delivery cost of dynamic multicast
D_{IV}	packet delivery cost of MBMS
$E[\cdot]$	expectation
E_b	energy per bit
H	number of required parity packets per TG for hybrid ARQ
I_{oc}	intercell interference power
I_{sc}	intracell interference power
I_0	interference power
K	number of PDUs per SDU
K_c	number of cells in hexagonal layout
L	number of retransmission attempts
L_j	propagation loss between MS and j BS
M	number of transmissions per received packet
M_B	average number of crossings for a cell per unit time
M_c	average number of crossings per unit time
M_{RA}	average number of crossings for an RA per unit time
M_{RNC}	average number of crossings for the area of an RNC per unit time
M_{SGSN}	average number of crossings for the area of an SGSN per unit time

M_{URA}	average number of crossings for a URA per unit time
M'	number of transmissions per received packet without accounting for parity packets
N_B	number of Node Bs
$N_{B/U}$	number of Node Bs per URA
$N_i^{(B)}$	number of cells of class i
$N_i^{(RA)}$	number of RAs per class i
$N_i^{(RNC)}$	number of RNCs of class i
N_m	number of members in multicast group
N_{RA}	number of RAs
N_{RNC}	number of RNCs
$N_{R/R}$	number of RNCs per RA
$N_{R/S}$	number of RAs per SGSN
$N_{S/G}$	number of SGSNs connected to a single GGSN
N_t	total number of subscribers in the network
$N_{U/R}$	number of URAs per RNC
N_{URA}	number of URAs
N_0	power spectral density of backgroud noise
P	error process transition probability matrix
P_{act}	probability that subscriber is SM active
P_b	base station outage probability
P_c	PTM channel outage probability
P_{cell}	probability that subscriber is cell connected
P_{conn}	probability that subscriber is PMM-CONNECTED
P_{det}	probability that subscriber is PMM-DETACHED
P_{in}	probability that subscriber is SM inactive
P_j	total transmitted power from jth BS
P_{RA}	probability that subscriber is PMM-IDLE
$P_{S/G}$	probability that SGSN does not have any RA serving multicast users
P_{URA}	probability that subscriber is URA connected
$Q(x)$	Q-function
R	data rate
R_c	cell radius
$R_{\max}$	maximum number of packet errors in TG for all receivers
S_b	cost of broadcasting a single packet using control signalling
S_{bu}	transmission cost of signalling over the air interface
S_{gs}	transmission cost of signalling between GGSN and SGSN
S_{hg}	transmission cost of signalling between HLR and GGSN
S_j	received power from jth BS
S_{rb}	transmission cost of signalling between RNC and Node B
S_{sr}	transmission cost of signalling between SGSN and RNC
U_{cell}	cell update cost
$U_{RA}^{(c)}$	RA update cost for PMM-CONNECTED users

$U_{RA}^{(i)}$	RA update cost for PMM-IDLE users
U_{RNC}	cost of intra-SGSN RNC relocation procedure
$U_{SGSN}^{(c)}$	SGSN update cost for PMM-CONNECTED users
$U_{SGSN}^{(i)}$	SGSN update cost for PMM-IDLE users
U_t	total location update cost for dynamic multicast and multiple unicast
$U_t^{(c)}$	total location update cost for PMM-CONNECTED users
$U_t^{(i)}$	total location update cost for PMM-IDLE users
U_{URA}	URA update cost
U_{II}	total location update cost for MBMS
$U_{II}^{(c)}$	MBMS location update cost for PMM-CONNECTED users
$U_{II}^{(i)}$	MBMS location update cost for PMM-IDLE users
$Var[\cdot]$	variance
V_{RA}	cost of paging a subscriber that is PMM-IDLE
V_{URA}	cost of paging a subscriber that is URA-connected
V_{IV}	cost of session start/stop procedure for MBMS
W	spreading bandwidth
Z	lognormal random variable
$\tilde{Z}$	normal random variable
a	number of proactively transmitted parities in TG
a_b	processing cost of signalling at Node B
a_g	processing cost of signalling at GGSN
a_h	processing cost of signalling at HLR
a_r	processing cost of signalling at RNC
a_s	processing cost of signalling at SGSN
b_ε	average burst length of PDU error process
c_b	unit processing cost at Node B
c_g	unit processing cost at GGSN
c_h	unit processing cost at HLR
c_r	unit processing cost at RNC
c_s	unit processing cost at SGSN
d	distance
g	fraction of users that are in soft handover
h	total number of parity packets in FEC block
h_{bu}	unit transmission cost over the air interface
h_{gs}	unit transmission cost between GGSN and SGSN
h_{hg}	unit transmission cost between HLR and GGSN
h_{rb}	unit transmission cost between RNC and Node B
h_{sr}	unit transmission cost between SGSN and RNC
i	whole number
k	number of data packets in TG
l	perimeter
l_B	cell perimeter

l_{gs}	number of packet hops between GGSN and SGSN
l_{hg}	number of packet hops between HLR and GGSN
l_{RA}	perimeter of RA
l_{rb}	number of packet hops between RNC and Node B
l_{RNC}	perimeter of RNC
l_{SGSN}	perimeter of SGSN
l_{sr}	number of packet hops between SGSN and RNC
l_{URA}	perimeter of URA
m	number of PDUs in round-trip time
n	number of packets in TG
n_B	number of cells with multicast users
n_c	number of children of node
n_i	number of MUCs per MGC
n_{RA}	total number of RAs that have multicast users
$n_{S/G}$	total number of SGSNs that are serving multicast users
p	packet loss rate
p_b	processing cost of packet delivery at Node B
p_g	processing cost of packet delivery at GGSN
p_n	packet loss rate of node
p_r	processing cost of packet delivery at RNC
p_s	processing cost of packet delivery at SGSN
q	residual packet loss rate of link layer
r_i	number of required parity packets at receiver i
v	velocity
w_{dt}	weighting factor of transmission cost for packet transmissions
w_p	weighting factor of processing cost
w_{st}	weighting factor of transmission cost for signalling
w_t	weighting factor of transmission cost
y	lognormal random variable
z	lognormal random variable
α	proportion of class i RAs
α_m	duty cycle for multicast
α_v	duty cycle for voice
γ	required E_b/I_0
δ	multicast population of class i RAs
ε	PDU packet loss rate
ε_{eff}	effective packet loss rate for multicast
ε_{res}	residual PDU error rate
η	average downlink power factor
$\theta_i^{(B)}$	number of multicast users for cell of class i
$\theta_i^{(RA)}$	number of multicast users for RA of class i
θ_1	multicast user population of class 1 RA
θ_2	multicast user population of class 2 RA
λ	arrival rate

ξ	normal random variable
ρ	population density per unit area
σ	standard deviation of shadowing
σ_e	spread of power control error
τ	throughput
φ	orthogonality factor
ϕ_i	fraction of total BS transmit power for ith MS
χ	lognormal random variable
ψ	fraction of total transmit power allocated for traffic
ω	path loss exponent

1

Introduction

This book focuses on the deployment of multicast in third-generation networks. Multicast is the efficient delivery of data to a group of destinations simultaneously. With multicast, messages are delivered as much as possible only once over each link of the network, creating copies only when the links to the destinations split.

In this chapter, we firstly provide an introduction to cellular mobile communication systems, in particular with respect to the features that distinguish the different generations of mobile communication systems, from analog first-generation to the fourth-generation systems currently in development. Then, we describe several fundamental aspects of data networking that are relevant for multicast. This is followed by an overview of how multicast can be achieved in data networks. We then introduce the basics of Internet Protocol (IP) multicast, the standard for multicast in internetworks. A more detailed description of IP multicast is provided in Chapter 2. Finally, we describe several existing mechanisms for carrying out multicast in third-generation networks. Several of these multicast mechanisms are described in much more detail in later chapters.

1.1 Cellular Mobile Communication Systems

The mobile communications industry is a relatively young industry. The basic technological concept of the industry lies in using radio waves to transmit data and connect users. Radio is the transmission of signals by modulation of electromagnetic waves with frequencies below those of visible light. Electromagnetic radiation travels by means of oscillating electromagnetic fields that pass through the air and the vacuum of space. Information is carried by systematically changing or modulating some property of the radiated waves, such as amplitude, frequency or phase. When radio waves pass an electrical conductor, the oscillating fields induce an alternating current in the conductor. This can be detected and transformed into sound or other signals that carry information.

The concept of using radio waves for communication dates back to the second half of the nineteenth century, when the German scientist Heinrich Rudolf Hertz demonstrated in 1888 that an electric spark of sufficient intensity at the emitting end could be captured by an appropriately designed receiver and induce action at a distance. This proved for the first time that electromagnetic waves propagate through the air and have the same properties as light. His English forerunner James Clark Maxwell had foreseen this a few

Multicast in Third-Generation Mobile Networks   Robert Rümmler, Alexander Gluhak and A. Hamid Aghvami
© 2009 John Wiley & Sons, Ltd

years earlier in 1864. Maxwell's theory of electromagnetic fields claimed the existence of electromagnetic waves and presented four mathematical formulae known today as Maxwell's equations, a set of fundamental equations governing electromagnetism.

Nikola Tesla first demonstrated the feasibility of wireless communications in 1893. He holds the US patent for the invention of the radio, defined as the *wireless transmission of data*. Guglielmo Marconi demonstrated the use of radio for wireless communications by equipping ships with life-saving wireless communications and by establishing the first commercial transatlantic radio service in 1907. Today, the use of radio takes many forms, including wireless and mobile communication of all types, as well as radio broadcasting.

1.1.1 The Cellular Concept

The design objective of early mobile communication systems was to achieve a large coverage area by using a single, high-power transmitter with an antenna mounted on a tall tower, transmitting on a single frequency. While this approach achieved very good coverage, it also meant that it was impossible to reuse the same frequency throughout the system, since any attempts to achieve frequency reuse would result in interference.

The cellular concept was a major breakthrough in solving the problem of spectral congestion and user capacity. It offered very high capacity with limited spectrum without any major technological changes. The cellular concept is a system-level idea that calls for replacing a single, high-power transmitter with many low-power transmitters, each providing coverage to only a small portion of the service area, referred to as a cell. Each base station is allocated a portion of the total number of channels available to the entire system, and nearby base stations are assigned different groups of channels so that all the available channels are assigned to a relatively small number of neighbouring base stations.

The mobile transceivers (also referred to as mobile phones, mobile stations, mobile terminals, handsets or devices) exchange radio signals with any number of base stations. Mobile phones are not attached to a particular base station, but may make use of any one of the base stations provided by the company that operates the corresponding network. The ensemble of base stations covers the landscape in such a way that the user can travel around and carry on a phone call without interruption, possibly making use of more than one base station. The procedure of changing a base station at cell boundaries is called *handover*.

Communication from the Mobile Station (MS) to the Base Station (BS) takes place on the uplink channel or reverse link, and from BS to MS on the downlink channel or forward link. To sustain a bidirectional commmunication between a mobile terminal and a base station, transmission resources must be provided both in the uplink and downlink directions. This can happen either through Frequency-Division Duplex (FDD), whereby uplink and downlink channels are assigned on separate frequencies, or through Time-Division Duplex (TDD), where uplink and downlink transmissions occur on the same frequency, but alternate in time.

FDD is efficient in the case of symmetric traffic. Also, FDD makes radio planning easier and more efficient, since base stations do not interfere with each other as they transmit and receive in different sub-bands. TDD has a strong advantage in the case where the asymmetry of the uplink and downlink data speed is variable. As the amount of uplink data increases, more bandwidth can dynamically be allocated to that, and as it shrinks it can be taken away. Another advantage is that the uplink and downlink radio paths are likely to be very similar in the case of a slow-moving system.

1.1.2 Propagation Impairments in Cellular Systems

The design of cellular systems is particularly challenging because of the adverse propagation conditions of the radio channel. Three main propagation impairments are usually distinguished. These are pathloss, slow fading or shadowing and fast fading or multipath fading (Brand and Aghvami, 2002).

The pathloss describes the average signal attenuation as a function of the distance between transmitter and receiver, which includes the free-space attenuation as one component, but also other factors come into play in cellular communications, resulting in an environment-dependent pathloss behaviour. Shadowing or slow fading describes slow signal fluctuations, which are typically caused by large structures, such as big buildings, obstructing the propagation paths. Fast or multipath fading is caused by the fact that signals propagate from transmitter to receiver through multiple paths, which can add at the receiver constructively or destructively, depending on the relative signal phases. The received signal is said to be in a deep fade when the paths add destructively such that the received signal level is close to zero. Fades occur roughly once every half-wavelength (Steele and Hanzo, 1999). With wavelengths of 30 cm and less in cellular communication systems, it is clear that multipath fading can result in relatively fast signal fluctuations (Brand and Aghvami, 2002).

1.1.3 Multiple-Access Schemes

Multiple-access schemes allow several devices connected to the same physical medium to transmit over it and to share its capacity. A multiple-access scheme is based on a multiplex method that allows several data streams or signals to share the same communication channel or physical media. Multiplexing is a term used to refer to a process where multiple data streams are combined into one signal over a shared medium. The resources that may be allocated with a multiple-access scheme are frequency bands, time slots, sets of codes or any combination of the three. The basic multiple-access schemes are Frequency-Division Multiple Access (FDMA), Time-Division Multiple Access (TDMA) and Code-Division Multiple Access (CDMA).

In FDMA, each communication is carried over one or two (depending on the duplexing method) narrowband frequency channels. The channel bandwidth and the modulation scheme determine the gross bit rate that can be sustained. With non-ideal filters, guard bands must be introduced between the FDMA channels to avoid so-called adjacent channel interference.

In TDMA, rather than assigning each user a channel with its own frequency, users share a channel of a wider bandwith in the time domain. This is achieved by introducing a framing structure, with each TDMA frame subdivided into a number of slots equal to the number of users that are to be supported. Provided that enough spectrum is available, multiple carriers may be assigned to each cell. Therefore, such TDMA systems typically feature also an FDMA element and are thus in reality hybrid TDMA/FDMA systems (Brand and Aghvami, 2002). TDMA systems must carefully synchronize the transmission times of all the users to ensure that they are received in the correct slot and do not cause interference.

In CDMA, narrowband signals are transformed through spectrum spreading into signals with a wider bandwidth. As in TDMA, multiple users share the carrier bandwidth, but, as in FDMA, they transmit continuously during the connection. The multiple-access capability derives from the use of different spreading codes for individual users. Because of the spreading of the spectrum, CDMA systems are also referred to as spread-spectrum multiple-access systems. Two basic CDMA techniques suitable for mobile communications are distinguished, namely Frequency-Hopping (FH) and Direct-Sequence (DS) CDMA techniques.

In an FH-CDMA system, a transmitter *hops* between available frequencies according to a specified algorithm, which can be either random or preplanned. The transmitter operates in synchronization with a receiver, which remains tuned to the same centre frequency as the transmitter. A short burst of data is transmitted on a narrowband carrier. Then, the transmitter tunes to another frequency and transmits again. Thus, the receiver is capable of hopping its frequency over a given bandwidth several times a second, transmitting on one frequency for a certain period of time, then hopping to another frequency and transmitting again. Frequency hopping requires a much wider bandwidth than is needed to transmit the same information using only one carrier frequency.

In a DS-CDMA system, a bit stream is multiplied by a direct sequence or spreading code composed of individual chips. They have a much shorter duration than the bits of the user bit stream, and this is why the original signal's spectrum is spread. The bandwidth expansion factor or spreading factor that results from using a transmission bandwidth that is several orders of magnitude greater than the minimum required signal bandwidth is equal to the duration of a bit divided by the duration of a chip.

1.1.4 First- and Second-Generation Systems

Various first-generation cellular mobile communication systems were introduced in the late 1970s and early 1980s. These early systems were characterized by analog (frequency modulation) voice transmission and limited flexibility. The first such system, the Advanced Mobile Phone System (AMPS), was introduced in the US in the late 1970s. Other first-generation systems include the Nordic Mobile Telephone (NMT) and the Total Access Communication System (TACS). The former was introduced in 1981 in Sweden, then soon afterwards in other Scandinavian countries, followed by the Netherlands, Switzerland and a large number of Central and Eastern European countries. The latter was deployed from 1985 in Ireland, Italy, Spain and the UK (Brand and Aghvami, 2002).

While these systems offered reasonably good voice quality, they provided limited spectral efficiency. They also suffered from the fact that network control messages – for handover or power control, for example – are carried over the voice channel in such a way that they interrupt speech transmission and produce audible clicks, which limits the network control capacity (Goodman, 1990).

The breakthrough of mobile telephony into the mass market occurred only in the 1990s with the advent of digital technology and the introduction of second-generation systems. Capacity increase was one of the main motivations for introducing second-generation systems. With digital technology it became possible to increase capacity by relying on low-bit-rate speech codecs and also integrating voice and data. Also, security was improved, both by means of encryption to provide privacy and authentication to prevent unauthorized access and use of the system. Dedicated channels were used for the exchange of network control information between mobile terminals and the network infrastructure during a call in order to overcome the limitation in network control of first-generation systems.

The Global System for Mobile Communications (GSM) is currently the uncontested standard for second-generation digital cellular communications. GSM, a TDMA-based system with optional slow frequency hopping, has a footprint covering virtually every angle of the world. With a subscriber number close to 500 million and a share of the digital cellular market close to 70 % in early 2001 (Brand and Aghvami, 2002), GSM is truly *the* global system for mobile communications. The General Packet Radio Service (GPRS) is a best-effort packet-switched service, which was designed as an enhancement to existing GSM networks in order to support non-real-time packet data traffic.

In the US, there are essentially two types of second-generation cellular system that are incompatible with each other. The first is a TDMA system called North American Digital Cellular or Digital AMPS (D-AMPS) and referred to as TDMA. The second system, which was launched later, is cdmaOne, the first operational CDMA system (Goodman, 1990). The relevant air interface specifications are the so-called interim standards IS-136 (for D-AMPS) and IS-95 (for cdmaOne).

The first and most popular Japanese second-generation standard is Personal Digital Cellular (PDC). It was later complemented by the Personal Handphone System (PHS), a mixture between mobile and cordless systems, which caters for low mobility, but is popular for certain applications owing to its relatively high data rates of up to 64 kilobits per second (kbps). Both standards are TDMA-based and have not seen wide deployment outside Japan (Brand and Aghvami, 2002).

1.1.5 Third-Generation Systems

Third-generation mobile networks represent the latest phase in the evolution of cellular technology, following from the first-generation analog and second-generation digital systems. Third-generation systems represent a shift from voice-centric services to converged services, including voice, data, video and so forth. In order to allow for advanced services and applications, third-generation networks provide higher capacity and enhanced network functionality.

Already before the launch of second-generation systems, the research community started to think about requirements for a new, third generation of mobile communication systems and about possible technological solutions to meet them. The European Telecommunication Standards Institute (ETSI) was one of the major players regarding the standardization of third-generation systems. It called its third-generation representative Universal Mobile Telecommunications System (UMTS) and established a number of requirements, according to which such a system should be designed.

The International Telecommunications Union (ITU) initially had the intention of controlling the standardization process such that a single system would emerge. With several bodies submitting their proposals for third-generation systems to the ITU in 1998, it soon became clear that the ITU would not be in a position to enforce a unified system. As a result of this, the ITU then advocated the concept of a *family of systems*, defined as a federation of systems referred to as International Mobile Telecommunications

2000 (IMT-2000). Eventually, two main camps formed. The first one is united in the Third-Generation Partnership Project (3GPP), dealing with the standardization of UMTS and the evolution of GSM, and the second one in a similar structure, the Third-Generation Partnership Project 2 (3GPP2), dealing with CDMA2000, an evolution of cdmaOne. The following sections provide a brief overview of third-generation systems.

Third-Generation Operating Modes

The ITU defines a third-generation network as one that delivers, among other capabilities, improved system capacity and spectrum efficiency compared with second-generation systems. The ITU's definition of third-generation technology stipulates, among other things, that these must be capable of supporting data transmission speeds of at least 144 kbps outdoors and 2 Mbps in fixed (indoor) environments (Zanoio and Urvik, 2003).

The IMT-2000 standard of the ITU consists of five operating modes, including three based on CDMA technology. The third-generation modes based on CDMA are most commonly known as Wideband CDMA (WCDMA), CDMA2000 and Time-Division Synchronous CDMA (TD-SCDMA), all of which are single-carrier DS-CDMA air interfaces.

WCDMA is the air interface used in UMTS networks. WCDMA is a wideband DS-CDMA air interface that achieves higher speeds and supports more users compared with the implementation of TDMA used in GSM networks. WCDMA, which operates with a carrier spacing of 5 MHz, can theoretically offer data transmission speeds of up to 2 Mbps.

CDMA2000 is a direct successor to IS-95 or cdmaOne. The CDMA2000 standards are CDMA2000 1xRTT, CDMA2000 Evolution Data Only (EV-DO) and CDMA2000 Evolution Data Voice (EV-DV). The CDMA2000 standard has evolved continually to support new services in a standard 1.25 MHz carrier. CDMA2000 1xRTT, with RTT standing for Radio Transmission Technology, provides data transmission capability with peak data at around 144 kbps. The data rate and cell capacity can be increased using the EV-DO and EV-DV. The data-only system CDMA2000 EV-DO offers theoretical speeds of up to 2.4 Mbps. EV-DV offers both data and voice, but has yet to be deployed anywhere.

TD-SCDMA is the mobile telecommunications standard being pursued in China. TD-SCDMA was incorporated by the 3GPP as part of UMTS Release 4. TD-SCDMA operates with a channel bandwidth of 1.6 MHz and offers theoretical data rates of up to 2 Mbps. TD-SCDMA uses TDD, in contrast to the FDD scheme used by WCDMA. Using the same carrier frequency for uplink and downlink means that the channel condition is the same on both directions, and the base station can deduce the downlink channel information from uplink channel estimates, which is helpful to the application of beamforming techniques. TD-SCDMA is being promoted as China's own third-generation solution, with only a limited number of vendors to offer TD-SCDMA technology. Commercialization of TD-SCDMA lags behind WCDMA by roughly 2 years, with TD-SCDMA networks having yet to be deployed (Tuttlebee and Payne, 2004/2005).

The other two options that are part of ITU's third-generation definition are Enhanced Data Rates for GSM Evolution (EDGE), a digital mobile phone technology that allows increased data transmission rates and improved data transmission reliability compared with GSM, and Digital Enhanced Cordless Telecommunications (DECT), a standard for digital portable phones, commonly used for domestic or corporate purposes.

CDMA2000 and EDGE are considered as *evolutionary* standards in that they can operate within existing spectrum allocations. WCDMA, TD-SCDMA and DECT, on the other hand, are *revolutionary* standards in that they require new spectrum allocation.

Deployment of Third-Generation Systems

In many European countries, a limited number of licences for radio spectrum required to operate WCDMA was allocated in a sealed bid auction. The license fees for spectrum were very high in some European countries, especially in the UK and Germany, in part due to the initial excitement over the potential of third-generation networks. Roll-out of third-generation networks was delayed in these countries because of this. Additional delays also resulted from the high cost of upgrading equipment for the new systems.

In Europe, the first commercial third-generation services were introduced in the UK and Italy, starting in March 2003. As of June 2007, 200 million third-generation network subscribers had been connected (Wikipedia, 2008a). This accounts for roughly 6.7 % of the 3 billion mobile phone subscriptions worldwide. In the countries where third-generation networks were launched first, namely Japan and South Korea, over half of all subscribers use third-generation networks.

As of December 2007, 190 third-generation networks were operating in 40 countries (The Global Mobile Suppliers Association, 2008). In Europe, the leading country is Italy, with a third of its subscribers migrated to third-generation networks. Other leading countries include the UK, Austria, Australia and Singapore at the 20 % migration level.

1.1.6 Towards Fourth-Generation Systems

It is predicted that, between 2012 and 2015, the current and future evolutions of third-generation networks will not have sufficient capacity to meet traffic demands. Regulatory and standardization bodies are therefore working towards commercial deployment of fourth-generation networks in that timeframe. Fourth-generation systems will be able to provide a comprehensive IP solution based on IPv6, where voice, data and streamed multimedia can be given to users on an *anytime, anywhere* basis, and at higher data rates than previous generations.

It will clearly be difficult to define the dividing line between third-generation and fourth-generation technology. A number of so-called fourth-generation technologies such as Long-Term Evolution (LTE) from 3GPP and Ultra Mobile Broadband (UMB) from 3GPP2 are in fact actually evolutions of third-generation technologies.

3GPP LTE is the name given to a project within the 3GPP to improve the UMTS standard to cope with future technology evolutions. Goals include improving spectral efficiency, lowering costs, improving services, making use of new spectrum and refarmed spectrum opportunities, and better integration with other open standards. The LTE project is not a standard, but it will result in the new evolved Release 8 of the 3GPP specifications, including mostly or wholly extensions and modifications of the UMTS system. UMB is the next-generation 3GPP2 standard air interface. UMB will use Orthogonal Frequency-Division Multiplexing (OFDM) rather than CDMA.

One of the drivers for the popular use of fourth-generation technology has been the aggressive promotion within the industry of the Institute of Electrical and Electronics Engineers (IEEE) 802.16e or WiMax mobile standard. A version of this standard was, however, recently accepted by the ITU as an addition to the IMT-2000 family, and therefore is clearly to be considered together with the other third-generation IMT-2000 technologies.

1.2   Networks and Protocols

A communications network is an interconnected group of computers or devices that have the ability to exchange data. The basic function of any network is to deliver data from a source to one or more receivers. Examples of networks are Local-Area Networks (LANs), which are constrained to small geographic areas, and Wide-Area Networks (WANs), which cover larger geographic areas than LANs. An immense number of technologies have been developed to accomplish this task, including a wide variety of physical media for transmitting data, such as twisted-pair copper wire cable or optical fibre, different ways for organizing network devices, as well as many protocols to govern how everything fits together to form a functioning whole (Kosiur, 1998).

Protocols are the rules that determine how a network operates. Protocols have two important roles. Firstly, they describe the syntax and semantics of messages exchanged within the network, that is, in what format messages are transmitted and what meaning these messages have. Secondly, protocols also define the actions that should be taken upon receipt of a message (Kosiur, 1998).

The Open Systems Interconnection (OSI) reference model is a layered, abstract description for communications and computer network protocol design. It was developed as part of the OSI initiative and is sometimes known as the OSI seven-layer model. As depicted in Figure 1.1, the OSI reference model consists of the application, presentation, session, transport, network, data link and physical layers. A layer is a collection of related functions that provides services to the layer above it and receives service from the layer below it. For example, a layer that provides error-free communications across a network provides the path needed by applications above it, while it calls the next lower layer to send and receive packets.

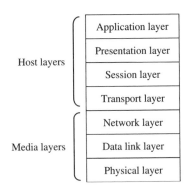

Figure 1.1   Layers of the OSI reference model

The layers in the OSI reference model are briefly described in the following (Zimmermann, 1980):

1. The **application layer** performs application services for the application processes and issues requests to the presentation layer. The application layer provides services to user-defined application processes, and not to the end-user itself.
2. The **presentation layer** establishes a context between application-layer entities, in which the higher-layer entities can use different syntax and semantics, as long as the presentation layer understands both and the mapping between them.
3. The **session layer** controls the dialogues/connections (sessions) between computers. It establishes, manages and terminates the connections between the local and remote application.
4. The **transport layer** provides transparent transfer of data between end-users, providing reliable data transfer services to the upper layers. The transport layer controls the reliability of a given link through flow control, segmentation/desegmentation and error control. Some protocols are state and connection oriented. This means that the transport layer can keep track of the segments and retransmit those that fail.
5. The **network layer** provides the functional and procedural means of transferring variable-length data sequences from a source to a destination via one or more networks. The network layer performs network routing functions, and might also perform fragmentation and reassembly, and report delivery errors.
6. The **data link layer** provides the functional and procedural means to transfer data between network entities and to detect and possibly correct errors that may occur in the physical layer.
7. The **physical layer** defines all the electrical and physical specifications for devices. In particular, it defines the relationship between a device and a physical medium.

1.2.1 Circuit-Switched and Packet-Switched Networks

Networks can be categorized by how data is routed and under what conditions data is accepted by individual network devices. In circuit-switched networks, a physical path is obtained for and dedicated to a single connection between two endpoints in the network for the duration of the connection. In packet-switched networks, small units of data called packets are routed through the network based on the destination address contained within each packet. Circuit-switched networks are often called connection-oriented networks, while packet-switched networks are referred to as connectionless networks.

With circuit switching, the source and destination must establish a connection to exchange data. The advantage of a circuit-switched network is its guaranteed capacity. Once a circuit is established, no other network activity can decrease the circuit's capacity. The main disadvantage of circuit switching is the fixed cost of the circuit, which is independent of the amount of traffic that flows on it. Some examples of circuit-switched networks are the Public Switched Telephone Network (PSTN) and the Integrated Services Digital Network (ISDN).

With packet switching, data between two end-systems is not sent as a continuous stream of bits. Instead, it is divided into small units called packets that are sent one at a time. These packets are multiplexed or allocated different time slots for transmission. More than one source can inject packets over the same wires into packet networks, so that, when a source is not transmitting, network resources are available for use by other sources. To allow the network to sort out these multiple flows of data, each packet carries an identifier of its destination. Thus, logical paths instead of physical circuits exist between communicating end-systems. The main advantage of packet switching is that multiple communications among end-systems can occur concurrently. The disadvantage is that, as network activity increases, a given pair of communicating end-systems receives less of the network capacity. Some examples of packet-switched networks are Asychronous Transfer Mode (ATM) and X.25 networks.

1.2.2 Internet Protocol Suite

The inability of a single type of network to satisfy all communication requirements necessitates connecting different networks together to create internetworks. The largest such internetwork is the Internet, a worldwide, publicly accessible collection of interconnected computer networks that transmit data by packet switching using standard IP. It is a *network of networks* that consists of millions of smaller domestic, academic, business and government networks. The IP suite is a set of communication protocols that implement the protocol stack on which the Internet and most commercial networks run. The IP protocol suite features various other protocols on top of IP, for example transport protocols such as the Transmission Control Protocol (TCP) and the User Datagram Protocol (UDP). The IP protocol suite is often referred to as TCP/IP, as these are the two most important protocols in the suite. TCP/IP is generally described as having four abstraction layers. The IP suite uses encapsulation to provide abstraction of protocols and services. Generally, a protocol at a higher level uses a protocol at a lower level to help accomplish its aims. The TCP/IP model and related protocols are currently maintained by the Internet Engineering Task Force (IETF).

From lowest to highest, the layers of the TCP/IP protocol suite are the network access layer, the network or internetwork layer, the transport layer and the application layer. The four layers are depicted in Figure 1.2. While the TCP/IP protocols do not fit neatly into all seven layers of the OSI reference model, they provide all the necessary functionality for productive networking.

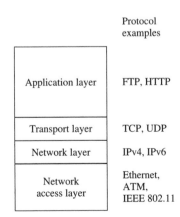

Figure 1.2 Layers of the TCP/IP protocol suite

The layers in the TCP/IP protocol suite perform the following functions (Kosiur, 1998):

1. The **application layer** organizes the data that the network will transfer. Examples of application protocols are the File Transfer Protocol (FTP) used to transfer data from one computer to another through a network and the Hypertext Transfer Protocol (HTTP) used to transfer information on the Intranet and the World Wide Web (WWW).
2. The **transport layer** is responsible for delivering the applications' information to the destination. Examples of transport protocols are TCP and UDP. TCP is a protocol that provides reliable, in-order delivery of a stream of bytes, making it suitable for applications such as file transfer and email. UDP, on the other hand, is a transport protocol that does not guarantee reliability or ordering in the way that TCP does. Packets may arrive out of order, appear duplicated or go missing without notice. Avoiding the overhead of checking whether every packet has actually arrived makes UDP faster and more efficient for applications that do not need guaranteed delivery.
3. The **network layer** or internetwork layer bears the responsibility of understanding the topology of the network and forwarding information through the network to its destination. IP version 4 (IPv4) and IP version 6 (IPv6) are the only standard internetwork-layer protocols used on the Internet. In order to forward data to the appropriate destination, network protocols rely on an addressing scheme for the hosts that are to receive the data. IPv4 is the first version of the protocol to be widely deployed. IPv6 is designated as the successor of IPv4. One of the main changes brought by IPv6 compared with IPv4 is a much larger IP address space that allows for greater flexibility in assigning addresses.
4. The **network access layer** combines the data link and physical layer of the network technology. Examples of network access technologies in widespread use are Ethernet, ATM and IEEE 802.11, the set of standards for Wireless LAN (WLAN) computer communication, developed by the IEEE in the 5 GHz and 2.4 GHz public spectrum bands.

1.2.3   Routing in Internetworks

Routers and switches are decision points in a network – a decision is made as to where in the network packets should be forwarded. The difference between the two types of device is what part of a packet is scanned to make a forwarding decision. In order to make a forwarding decision, switches look at Medium Access Control (MAC) addresses in layer 2 of the OSI reference model protocol stack, whereas routers review the network address found in layer 3 of the protocol stack, the network layer.

A routing protocol helps routers to create a topological map of the network. Routing protocols are part of the internetwork layer of the TCP/IP protocol stack. Routing protocols are different in how they share information and compute routes. With link-state routing protocols, every node constructs a map of the connectivity of the network in the form of a graph showing which nodes are connected to which other nodes. Routers share only the identity of their neighbours, but they flood the entire network with this information. Distance-vector routing protocols periodically share their knowledge of the entire network, but only with their neighbours. As neighbouring routers learn new information, they pass that information on to their neighbours until, slowly but surely, the information makes its way across the entire network. With both types of routing protocol, routing tables are constructed, which are held in the router's memory. The routing process forwards packets on the basis of routing tables which maintain a record of the routes to various network destinations.

More specifically, link-state routing protocols create the map in three phases: each router meets its neighbours (learn your neighbourhood), routers then share that information with other routers on the network (learn about other neighbourhoods) and, finally, routers combine the information and calculate routes (Kosiur, 1998). Link-state protocols flood the entire network with link-state information, which can lead to unnecessary bandwith usage. This disadvantage is offset by the speed and efficiency of link-state routing. The best known link-state routing protocol is the Open Shortest Path First (OSPF) routing protocol.

1.3   Multipoint Communications

Point-to-multipoint or more simply multipoint communications is a term used in the telecommunications field to refer to communication that is accomplished via a multi-point connection, providing a communication link between a single source and multiple receivers.

Many applications involve one-to-many or many-to-many communications, where one or more sources are sending data to multiple receivers. Such multipoint transmissions can be achieved in three different ways: unicast where a separate copy of the data is delivered to each recipient, broadcast where a data packet is forwarded to all portions of the network and multicast where a single packet is addressed to all intended recipients and the network replicates packets only as needed.

1.3.1   Unicast

In networking, the fundamental method of communication is between two hosts, or unicas-ting. Such one-to-one sessions offer a great deal of control of the data traffic between the source and receiver, allowing for acknowledgement of receipt, requests for retransmission of data, changes in transmission rate and so on. Figure 1.3 illustrates the point-to-point nature of unicast.

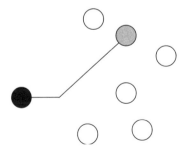

Figure 1.3   Unicast as one-to-one communication

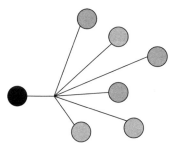

Figure 1.4   Broadcast as one-to-all communication

In multipoint unicasting, a source sends an individual copy of a message to each recipient. If five individuals in a workgroup wish to receive a copy of the same file using FTP, the server would send the file to each of the five recipients separately, using 5 times as much bandwidth as for a single transfer. This is wasteful of network bandwidth.

1.3.2 Broadcast

Under certain circumstances, it can be more efficient to transmit one copy of a message to all network nodes and let the receiving nodes decide if they want the message. Distributing the task of duplicating the packets among the network nodes rather than focusing the task at the sender's host machine is advantageous for the sender. This is referred to as broadcasting. Figure 1.4 illustrates the one-to-all nature of broadcast. There are many network hardware technologies that include mechanisms to broadcast packets to multiple destinations at the same time. With Ethernet, broadcast delivery can be accomplished with a single packet transmission on the wire.

One significant feature of broadcasting is that it relieves the source from the task of duplicating packets that are destined for multiple recipients. The source transmits a single copy of the packet to the appropriate broadcast address and the network devices take over, duplicating the packet as needed to cover the network.

1.3.3 Multicast

Multicast is the delivery of information to a group of destinations simultaneously using the most efficient strategy to deliver the messages over each link of the network only once, creating copies only when the links to the destinations diverge. Multicast falls between unicast and broadcast. Rather than sending data to a single host or to all hosts on the network, multicast aims to deliver data to a select group of hosts. The host group is defined by a specified multicast address. Figure 1.5 highlights the one-to-many nature of multicast.

Once a host group is set up and the sender starts transmitting packets to the host group address, the network infrastructure takes on the responsibility for delivering the necessary data streams to all members of the group. Only one copy of a multicast message passes over any link along the delivery path in the network. Copies of the message are only made when paths diverge at a router, thus helping to conserve bandwidth. In multicast, as in

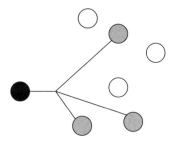

Figure 1.5   Multicast as one-to-many communication

broadcast, the source of a message usually does not know which recipients are within the group or the state of the data delivery.

On an LAN, each host's network interface monitors the LAN and accepts packets addressed to the multicast address that defines the host group to which the packets belong. Unlike broadcasting, multicasting allows each host to choose whether it wants to accept multicast packets. In a WAN, membership information concerning host groups has to be maintained across the entire WAN or internetwork. Procedures for joining a host group and maintaining a host group differ from the LAN case, since routers have to get involved, passing group information among themselves to maintain the required information for multicast within the internetwork.

1.4   IP Multicast

IP multicast is a technique for one-to-many communication over IP infrastructure. In IP multicast, routers create optimal distribution paths in real time for packets sent to a multicast destination address. IP multicast is designed to scale to a large receiver population by not requiring prior knowledge of who or how many receivers there are. Multicast utilizes network infrastructure efficiently by requiring the source to send a packet only once, even if it needs to be delivered to a large number of receivers. The nodes in the network take care of replicating the packet to reach multiple receivers only where necessary. In networking, the term *multicast* is synonomous with IP multicast. Stephen Deering was the first to describe the standard multicast model for IP networks (Deering, 1989, 1991; Deering and Cheriton, 1990). This model describes how end-systems are to send and receive multicast packets. The model includes both an explicit set of requirements and several implicit requirements.

We will briefly describe the key concepts of IP multicast, focusing on the minimum set of functionality required to provide multicast, namely group management and multicast routing. A much more detailed description of multicast group management and multicast routing, as well as other multicast mechanisms such as session announcement and reliable delivery of multicast traffic, is provided in Chapter 2.

1.4.1   Multicast Groups

The model for multicast builds on IP-style semantics as well as open and dynamic groups. A source can send multicast packets at any time, with no need to register, announce or schedule transmission. IP multicast is based on UDP (not TCP), so packets are delivered using a best-effort policy without reliability or congestion control. Sources only need to know a multicast address. They do not need to know group membership, and they do not need to be a member of the multicast group to which they are sending. A group can have any number of sources. Multicast group members can join or leave a multicast group at will. There is no need to register, synchronize or negotiate with a centralized group management entity.

An IP multicast group address is used by sources and the receivers to send and receive content. Sources use the group address as the IP destination address in their data packets. Receivers use this group address to inform the network that they are interested in receiving packets sent to that group. For example, if some content is associated with a specific multicast group address, the source will send data packets destined for the group to the multicast group address. Multicast receivers will inform the network that they are interested in receiving data packets sent to the multicast group. This is done by having the receiver *join* the multicast group. The protocol used by receivers to join a group is called the Internet Group Management Protocol (IGMP). IGMP is used by IP hosts and adjacent multicast routers to establish multicast group memberships. It is an integral part of the IP multicast specification, operating above the network layer, although it does not actually act as a transport protocol. The IGMP protocol is implemented on the host and router side. The host reports its membership of a group to its local router, with the router listening to reports from hosts and periodically sending out queries to check whether multicast group information has changed.

1.4.2 Multicast Routing

IP multicast does not require a source sending to a given group to know about the receivers of the group. In order to route multicast packets, the network constructs a multicast tree. The multicast tree construction is initiated by network nodes close to the receivers and is thus receiver-driven. This allows it to scale to a large receiver population. The multicast distribution tree is constructed for that group once the receivers join a particular IP multicast group.

There are several different multicast routing protocols, and each one has its own unique technological solution. The Distance-Vector Multicast Routing Protocol (DVMRP) is the earliest protocol for multicast routing. A key concept introduced by DVMRP is the use of separate forwarding trees for each multicast group. This fundamental principle continues to be used in the newer multicast routing protocols. DVMRP is a distance-vector protocol that provides very limited flexibility, functionality and scalability. Nonetheless, the early Internet multicast experiments in the multicast backbone (or MBone for short) were based on DVMRP. DVMRP has largely been superseded by some of the newer protocols.

The next incarnation of multicast routing was an extension of the popular OSPF protocol called MOSPF. OSPF is designed explicitly to be an interior gateway protocol, meaning that it resides within a single autonomous system. Hence, any extensions to OSPF, such as MOSPF, would also reside within the confines of one autonomous system. MOSPF provides an effective means for a single corporation, university or other organization to support multicast routing, but it cannot support wide-scale applications that require the use of the Internet. MOSPF is used sporadically for some specialized applications, but it is not prevalent.

A new breed of multicast routing protocols was developed in the late 1990s. This family of protocols is collectively known as Protocol-Independent Multicast (PIM). The name PIM is derived from the fact that these multicast forwarding protocols are not dependent upon any one specific routing protocol (MOSPF, for example, requires the use of the OSPF unicast routing protocol). Instead, PIM will take advantage of the existing routing tables, regardless of how they were constructed, in order to forward multicast data. There are a few different versions of PIM. One version is called PIM Dense Mode (PIM-DM). As the name implies, this particular rendering of PIM is optimized for densely populated communities of users. The philosophy of PIM-DM is to *push* multicast data towards the users. A router that deploys this protocol simply floods multicast traffic streams to all interfaces (this is similar to broadcast mechanisms). If the downstream routers are not attached to any users that require this particular multicast stream, they will send a *stop* message to the upstream router. This message is called a *prune* message, since the upstream router will then prune its forwarding tree to eliminate that particular branch. PIM-DM routers forward all multicast traffic until a downstream router objects. The most commonly implemented form of PIM is PIM Sparse Mode (PIM-SM). This is a direct contrast to PIM-DM since it invokes a *pull* methodology instead of a *push* technique. This means that PIM-SM routers must specifically request a particular multicast stream before the data is forwarded to them. PIM-SM is well suited for the Internet since it reduces the overhead and bandwidth requirements for multicast data streams.

1.5   Multicast in Cellular Mobile Networks

Only in recent years has resource-efficient multicast for third-generation networks received significant attention in the research community. The goal behind introducing multicast capabilities in third-generation networks is to realize cost efficiencies by transmitting the same data to multiple receivers on shared resources.

In UMTS, multicast can be achieved with the Cell Broadcast Service (CBS), IP multicast and the Multimedia Broadcast/Multicast Service (MBMS). CBS is a standard for GSM and UMTS that allows for simultaneous delivery of messages to multiple users in a specified area. IP multicast is standardized as an optional feature in UMTS networks, but is not efficient in terms of bandwidth consumption. MBMS, on the other hand, standardized by the 3GPP in Release 6, is designed to support efficient broadcast and multicast packet delivery in GPRS and UMTS networks. In CDMA2000 networks, multicast can be achieved by means of IP multicast or the Broadcast/Multicast Service (BCMCS). BCMCS provides multimedia content transmission from a single source to multiple users in CDMA2000 networks. The options for providing multicast in UMTS and CDMA2000 networks are briefly described below.

1.5.1   Cell Broadcast Service

With CBS, fairly small-sized messages are broadcast to all mobile handsets and similar devices within a designated geographical area. The broadcast range can be varied, from a single cell to the entire network. CBS messages originate from a Cell Broadcast Entity (CBE). CBEs are usually connected to a Content Casting Centre (CCC), which is in turn connected to the Cell Broadcast Centre (CBC). CBS messages are then sent from the CBC to the cells, in accordance with the CBS's coverage requirements.

A CBS message page comprises 82 octets, which, using the default character set, equates to 93 characters. Up to 15 of these pages may be concatenated to form a CBS message. Each page of such a message will have the same message identifier (indicating the source of the message) and the same serial number. Using this information, the mobile telephone is able to identify and ignore broadcasts of already received messages. CBS is not widely deployed today. In the USA, for example, most handsets do not have cell broadcast capabilities and the major network operators have not deployed the technology in their networks.

CBS messaging is particularly appropriate for emergency purposes, as it is not as affected by traffic load. It is therefore usable during a disaster when load spikes tend to crash networks. Taking the Tsunami catastrophe in Asia as an example, an operator in Sri Lanka was able to provide ongoing emergency information to its subscribers, to warn of incoming waves, to give news updates, to direct people to supply and distribution centres and even to arrange donation collections by relying on CBS (Wikipedia, 2008b).

1.5.2 IP Multicast

IP multicast is standardized as an optional feature in UMTS networks. With this feature, the IP multicast routing protocol is terminated at the gateway of the UMTS network. As a result, this solution requires that only the UMTS gateway be multicast aware. A similar mechanism applies to CDMA2000 networks. With IP multicast in UMTS, the gateway serves as an IGMP-designated router and performs IGMP signalling on point-to-point packet-data channels. Multicast data is forwarded to the receiver on point-to-point or unicast channels. Only the UMTS terminal and the gateway are multicast aware. This architecture allows the network to treat multicast traffic in the same manner as unicast traffic. With this architecture, however, no bandwidth savings can be achieved in the network. Multicast packets are duplicated at the gateway and transmitted to each multicast group member individually.

This multicast architecture reduces the load on a wireless source that wishes to transmit multicast traffic within the network. The source only needs to send one copy of multicast packet data to the gateway. The gateway is reponsible for forwarding multicast packets on to the multicast distribution tree. The drawback of this design is that the UMTS multicast source does not receive any information from multicast members. When the multicast group does not have any members, the source continues to transmit its multicast data to the gateway. The source is not aware of the empty state of the multicast group. A modified signalling connection between the gateway and the source can avoid this situation.

1.5.3 MBMS for UMTS

MBMS is a standard for multimedia broadcast and multicast content delivery in UMTS networks. MBMS has two operational modes, namely broadcast and multicast. In broadcast mode, transmission takes place regardless of user presence in a defined area, whereas, in multicast mode, solely users that belong to a multicast group are serviced. Consequently, MBMS in multicast mode requires group management, whereas broadcast mode does not.

MBMS classifies different types of user service according to the method of distribution. The three user service types are streaming, file download and carousel services. Streaming services provide a stream of continuous media, for instance audio and video traffic. File download services are used to deliver binary files that can only be consumed when the file is downloaded in its entirety. Carousel services finally provide content that is retransmitted periodically. Chapter 6 provides a much more detailed description of MBMS.

1.5.4 BCMCS for CDMA2000

BCMCS is a CDMA2000 standard that provides the capability to deliver broadcast and multicast services. As with MBMS, the BCMCS is designed for multimedia content transmission from a single source to multiple users. BCMCS can be used for broadcast services, in which all users within the broadcasting area can receive the information, and multicast services, in which only users that have subscribed to the service can receive the information.

IETF protocols are widely employed in BCMCS. This minimizes the number of new protocols required and maximizes the utilization of well-accepted standards. BCMCS is described in much more detail in Chapter 7.

1.6 Summary

In this chapter, the most important aspects of multicast in third-generation networks have been introduced. Firstly, an introduction to cellular mobile communication systems was given, from first-generation analog systems to third-generation systems and beyond. This was followed by an overview of the most relevant aspects of data networking, covering such aspects as circuit-switched and packet-switched networks, the Internet protocol suite and routing in internetworks. We then described the basics of multipoint communications in data networks. Multipoint communications can be achieved by means of multipoint unicast, broadcast or multicast. With multipoint unicast, packets destined for a group are duplicated and transmitted to group members individually. With broadcast, data is transmitted within a given geographic area to all members of the network. With multicast, data is transmitted to only those recipients that have requested to be members of a given group. We then briefly described the key fundamentals of IP multicast, the technique for one-to-many communication over IP infrastructure. This was then followed by a brief overview of how multicast can be achieved in third-generation networks.

The subsequent chapters will provide much more detail on several key topics. IP multicast is described in Chapter 2, an overview of the third-generation networks UMTS and CDMA2000 is given in Chapter 3, followed by a complete description of the multicast mechanisms for these networks in Chapters 4 to 6. Chapters 8 to 10 then analyse the performance of these multicast mechanisms. Chapter 11 finally looks beyond multicast in third-generation networks and investigates how multicast can be achieved in heterogeneous networks.

2

Fundamentals of IP Multicast

This chapter provides an introduction to the fundamentals of IP multicast. It assumes that the reader is familiar with the basics of IP networking and builds on this knowledge to explain the concepts and mechanisms of IP multicast. Readers who are familiar with the basics of IP multicast can skip this chapter and proceed to the next chapter, which provides an overview of third-generation networks.

2.1 Introduction

Since its introduction in the late 1980s by Deering (1989), the functions and protocols for IP multicast have evolved over time. A brief overview of the most important functions and protocols is provided in the following.

Multicast address management deals with the assignment and the scope of multicast addresses. The initial multicast service model allowed senders to send data addressed to any valid multicast address. If two sources sending in the same scope use the same multicast address, unintentional address collisions may occur. Multicast address management aims to avoid address conflicts within a domain or between different domains by coordinating the assignment of addresses.

Multicast service announcement and discovery handles the ways in which non-members of a multicast group can be informed about an ongoing multicast session or discover an ongoing session. It also handles how a source can notify its willingness to send data to a specific multicast group.

Multicast group management deals with the collection and maintenance of the multicast group membership, the set of receivers that are interested in receiving data sent to the same multicast group. Multicast group management protocols operate between hosts and the first-hop multicast router on a subnet. Using a group management protocol, receivers notify multicast routers about the IP multicast groups in which they are interested. Routers collect and maintain this membership information for each of their interfaces.

Multicast routing protocols ensure that efficient delivery paths from one or more multicast sources to all receivers are established. The collection of paths over which a multicast packet is sent is called a multicast delivery tree. Multicast routing protocols are responsible for the construction and maintenance of multicast delivery trees that connect multicast

Multicast in Third-Generation Mobile Networks   Robert Rümmler, Alexander Gluhak and A. Hamid Aghvami
© 2009 John Wiley & Sons, Ltd

group members and for the forwarding of multicast packets on the multicast delivery tree. For this, multicast routing protocols utilize the group membership information collected by group management protocols.

Reliable multicast transport protocols ensure the reliable delivery of multicast traffic to a potentially large receiver group in a scalable manner. Most current IP multicast applications operate on top of UDP, which only provides best-effort delivery guarantees. Applications such as file delivery, however, require reliable multicast delivery. Reliable multicast protocols rely on strategies such as Forward Error Correction (FEC) and retransmission schemes that rely on positive Acknowledgements (ACKs) or Negative Acknowledgements (NACKs) from the receiver population to recover from losses experienced during the transmission.

Flow and congestion control protocols address issues such as controlling the rate of the sender to prevent receivers being overwhelmed by the data transmission, ensuring fairness between multicast sessions and TCP traffic in case of congestion, as well as ensuring fairness of rate among heterogeneous receivers within a multicast session.

Multicast mobility deals with the challenges caused by the mobility of multicast sources and receivers with respect to their point of network attachment. The proposed solutions address challenges such as ensuring seamless continuity of data reception while receivers are mobile or limiting the impact of source and receiver mobility on the multicast routing tree.

To date, a large number of protocols have been proposed within the research community to achieve the required functionality for IP multicast. This chapter starts with a description of the two service models for IP multicast. It then introduces the basics of multicast addressing in IPv4 and IPv6 networks. Two elementary aspects of IP multicast, namely group management and IP multicast routing, are then described in detail. The characteristics of initial mechanisms for IP multicast are examined with respect to their suitability in dealing with different delivery requirements. The reliable delivery of multicast traffic in particular poses additional challenges compared with the delivery of unicast traffic. We then discuss the challenges and several approaches for flow and congestion control of multicast traffic. Finally, mechanisms and extensions that support multicast delivery to mobile hosts are presented, focusing mainly on extensions of Mobile IP. The chapter concludes with a summary highlighting the key points introduced in the chapter.

2.2   IP Multicast Service Models

The initial multicast model introduced by Deering (1989) describes how end-systems in an IP network are able to send and receive multicast traffic on their local subnetwork. The model is a simple end-system specification that lays the foundations for the so-called Any Source Multicast (ASM) service model. The ASM service model is an *open* service model. It allows any end-system on the Internet to send to or receive data from a multicast group, with no mechanisms that restrict access to doing so. Multicast groups are completely open, and sources and receivers are only connected by the notion of the IP multicast address. Sources do not need to be members of the multicast group in order to send data to the group. Sources can send multicast data to the group at any time, without the need for prior registration or scheduling. A multicast group can have any number of sources. A source is usually unable to prevent another source from sending to the same multicast group. Sources are usually unaware of any group membership details. Any receiver is free to become a member of a given multicast group at any time. Receivers can join and leave multicast groups at will. With the ASM service model, receivers use IGMP version 2 (IGMPv2) to join or to leave a multicast group (Fenner, 1997). Once a receiver has joined a multicast group, it receives data from all sources sending to the multicast group address, regardless of who is sending the data.

Although the ASM service model is relatively simple, its openness and lack of access control make it very difficult to manage. Diot *et al.* (2000) argue that the poor manageability of the ASM service model has hindered the widespread deployment of IP multicast in recent years. Besides the lack of access control, problems with multicast address allocation (Bhattacharyya, 2003) and inefficient handling of traffic from well-known sources have motivated the introduction of a new service model, namely the Source-Specific Multicast (SSM) service model (Holbrook and Cain, 2006).

The SSM service model replaces the notion of an IP multicast address as the only identifier of a multicast group with that of a so-called *channel*. A channel is characterized by its source address S and its destination address G. This channel abstraction ensures that, for each multicast channel, only a single sender and any number of receivers may exist. Two sources, S_1 and S_2, sending to the same multicast group address, G, are regarded as two separate channels. A receiver that subscribes to the channel (S_1, G) will only receive the traffic sent by source S_1, even though other sources may be sending data to the destination address G. This simplifies global address allocation, as each sender is now responsible for resolving potential address collisions of the different channels it may create. Unlike in ASM, access control is implicitly ensured, as no other sender is able to transmit on the same channel. SSM also simplifies multicast routing, as only source-based forwarding trees are required.

As in the ASM service model, a sender in SSM is unaware of its receivers. Receivers, however, need explicitly to know the source address of the sender in order to clearly identify a multicast channel. This makes SSM particularly well-suited to dissemination-style applications with one or more senders whose identities are known before the application begins (Holbrook and Cain, 2006). In order to support subscriptions of receivers, IGMP version 3 supports the use of multicast source filters (Cain *et al.*, 2002).

2.3 Multicast Addressing and Address Management

Addressing is an essential part of IP multicast. In IP multicast, a source initiates a session by sending data to a specific multicast destination address. Multicast data is only received by recipients that have previously expressed an interest in receiving data for the group. A multicast address thus identifies a set of hosts, or, more specifically, a set of interfaces, that are interested in receiving data sent to a particular multicast address.

Multicast addresses occupy a dedicated address range within the IP address space. This is required so that multicast addresses can be distinguished from unicast addresses at routers. Multicast addresses usually have different scopes such as link-local or global scope. As the number of multicast addresses is limited, careful coordination of multicast

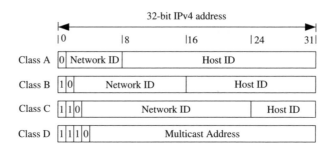

Figure 2.1 IPv4 address types

addresses in the different scopes is essential in order to avoid collisions within the multi-cast address space. Collisions usually occur if two unrelated sources start sending data to the same multicast address within the same scope. In order to avoid collisions, mechanisms for multicast address assignment are required.

This section gives an overview of addressing for IPv4 and IPv6 multicast, explains the mapping between IP multicast and IEEE 802 MAC addresses and describes existing solutions for multicast address allocation.

2.3.1 IPv4 Multicast Addressing

IPv4 addresses are represented by a 32-bit address identifier. The resulting IPv4 address space is partitioned into four different classes, as shown in Figure 2.1. Class A, B and C addresses are used for unicast and have, apart from an address prefix, a network and host component. Class D addresses are reserved for multicast and carry the multicast address as the sole identifier after the address prefix. Multicast addresses occupy the address range from 224.0.0.0 through to 239.255.255.255. The Internet Assigned Numbers Authority (IANA) is responsible for the assignment of IPv4 multicast addresses. IPv4 unicast addresses, on the other hand, are delegated in blocks to regional registries. Detailed guidelines on address assignment are given in (Albanna *et al.*, 2001).

2.3.2 IPv6 Multicast Addressing

In contrast to the relatively small number of available class D addresses in IPv4, the IPv6 address space provides a nearly unlimited amount of addresses. Figure 2.2 illustrates the format of an IPv6 multicast address. The first 8 bits of an IPv6 multicast address are always set to 1. Of the 128 bits of an IPv6 address, 112 bits define the group identifier (ID) used to identify a multicast group within a given scope. The scope of the group can vary from node-local, link-local, site-local, organization-local to global and is encoded in the 4 bits preceding the group IP. The transient flag identifies if the address has been assigned permanently to a multicast group or if it is a transient multicast address. As in the IPv4 address space, the IANA defines a set of permanently assigned multicast addresses within the IPv6 address space. These addresses are used for network control traffic and other applications. The main IPv6 multicast address assignments are given in (Hinden and Deering, 1998).

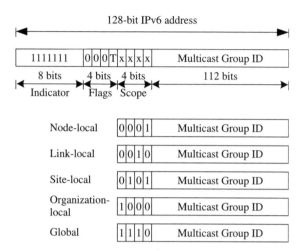

Figure 2.2   IPv6 multicast addresses

2.3.3   *Mapping between IP Multicast and MAC Addresses*

The hardware interface of a host usually filters packets based on the destination MAC address. Each IEEE 802-based network interface has a unique 48-bit MAC address. For IP unicast traffic, the IP address of a host is resolved to the hardware MAC address of a destination host via the Address Resolution Protocol (ARP) (Plummer, 1982).

In order to receive multicast traffic for a particular group, a receiver firstly has to notify its local multicast router of its interest to receive traffic destined for that group and then configure its interface to listen to the multicast address.

The hardware interface of a host usually provides the means for dynamic configuration of one or more multicast MAC addresses in order to enable filtering of multicast traffic on the local link. The highest-order 25 bits of the MAC address are fixed and common to any multicast address. The lowest 23 bits of the MAC address are variable and can be used for constructing different MAC addresses. In order to avoid explicit binding of IP addresses to MAC addresses, IPv4 multicast addresses can be mapped automatically to their corresponding MAC addresses. Figure 2.3 illustrates this mapping.

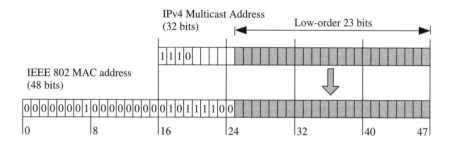

Figure 2.3   Mapping between IPv4 multicast and IEEE 802 MAC addresses

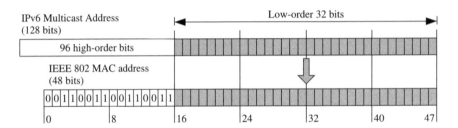

Figure 2.4   Mapping between IPv6 multicast and IEEE 802 MAC addresses

The lowest-order 23 bits of the IP multicast address are directly mapped to the corresponding bits of the MAC address. As the highest-order 4 bits are always the same for IP multicast addresses, 5 bits in the IP multicast address are left out of the mapping. As a result, the interface of a host may unnecessarily receive traffic of other multicast groups that have the lowest 23 bits in common. In practice, the unwanted traffic is discarded at the IP layer.

The mapping for IPv6 addresses is slightly different, as illustrated in Figure 2.4. The higher-order 16 bits of the MAC address are assigned a fixed value as an indicator of an IPv6 multicast address. The low-order 32 bits of the IPv6 multicast address are mapped directly to the remaining low-order 32 bits of the MAC address.

2.3.4   Assignment of Multicast Addresses

In order to facilitate a smooth multicast operation, it must be ensured that an IP multicast address is used uniquely within its scope. The IETF has proposed two different schemes to tackle address collisions. The Multicast Address Allocation Architecture (MAAA) provides multiple hierarchical server layers and a set of protocols for address allocation (Thaler, Handley and Estrin, 2000). The protocols handle the allocation of multicast addresses between different domains as well as within the domain of an operator. MAAA enables applications to obtain IP multicast addresses dynamically from allocation servers within an allocation domain, thus reducing the probability of address clashes to a minimum. Recently, the unicast-prefix-based multicast address mechanism has been proposed (Haberman and Thaler, 2002). It allows operators to assign multicast addresses without the need to run an interdomain multicast allocation protocol.

2.4   Multicast Session Announcement

Session announcements allow multicast receivers to discover the availability of multicast sources. The Session Announcement Protocol (SAP) handles multicast session announcements (Handley and Jacobson, 1998). SAP relies on two entities, namely the SAP announcer and the SAP listener. An SAP announcer periodically broadcasts announcement information in SAP packets carried over UDP and transmitted to a well-known multicast address. A multicast session is usually announced with the same scope as that of the multicast group. Depending on the scope, the SAP announcer must choose the correct announcement channel. The SAP listener entity listens to the well-known multicast address. This allows the SAP listener to become aware of any ongoing multicast sessions within the desired scope. The session announcement can carry various details regarding the session, such as a simple description of the session content, the session start and duration, special requirements needed to decode the session data and so on. The session information itself is carried as payload in the announcement packets. The content of the announcement packets is described by the Session Description Protocol (SDP) (Handley and Jacobson, 1998). SDP is used for describing multimedia sessions, session invitations and other forms of multimedia session initiation (Rosenberg *et al.*, 2002). SDP is purely a format for describing sessions and thus does not incorporate a transport protocol. The goal of SDP is to describe the relevant media components within a session in a way that allows the receiver to set up and join the session smoothly. SDP is also used in the Session Initiation Protocol (SIP) (Rosenberg *et al.*, 2002) and the Real-Time Streaming Protocol (RTSP) (Schulzrinne, Rao and Lanphier, 1998), as part of email that uses Multipurpose Internet Mail Extensions (MIME) and in HTTP.

An SDP session description consists of lines of text that take the form *type* = *value*, where *type* is a single case-sensitive character and *value* is a structured text string. When SDP is conveyed by SAP, only one session description is allowed per packet. The type parameter *v* specifies the version number of the SDP protocol. The owner of the session is described by the type parameter *o*. The subject of the session is described by the type parameter *s*, while the type parameter *i* provides additional textual information. The type parameter *c* finally provides the connection address, in this case the multicast address of the session. More information on SDP can be found in (Handley and Jacobson, 1998).

2.5   Group Management

Multicast data delivery is achieved by an interworking of local and global mechanisms. Local mechanisms are responsible for multicast group management, whereas global mechanisms are responsible for multicast routing. Using a multicast group management protocol, receivers are required to notify a multicast-enabled router within their subnetwork of their interest in receiving data for a particular multicast group. This process is referred to as joining a multicast group. Multicast routing protocols executed in routers use this information to establish appropriate multicast delivery paths to all receivers. Multicast data is then forwarded in a resource-efficient fashion along the multicast routing path to all receivers. Likewise, receivers can notify a router if they are not interested in receiving data for a particular multicast group. This process is referred to as leaving a multicast group.

A number of group management protocols have been standardized. IGMP is employed in IPv4 networks. Several versions of this protocol exist. Although IGMP version 3 (IGMPv3) is the most recent version, IGMPv2 is still the most widely deployed version of the protocol. The counterpart to IGMP in IPv6 networks is the Multicast Listener Discovery (MLD) protocol. Its functionality is identical to IGMP, with MLD version 1 (MLDv1) being equivalent to IGMPv2, and MLD version 2 (MLDv2) being equivalent to IGMPv3.

2.5.1   Internet Group Management Protocol

IGMP is an integral part of IPv4 and must be implemented by every host that wishes to receive multicast traffic. IGMPv2 messages are 8 octets long and are encapsulated in IP packets, with a protocol number of 2. Since IGMP messages should only reach routers directly attached to their local subnetwork, IP-encapsulated IGMP messages are sent with a Time-To-Live (TTL) value set to 1.

IGMP relies on a query-reply mechanism. Routers periodically query hosts for new membership information by means of group membership queries, with hosts replying to these queries by means of group membership reports. The interval between two queries can be configured but is typically 125 s. Besides periodic queries, which are referred to as general queries, so-called group-specific queries exist. While general queries can invoke membership reports for arbitrary multicast groups, group-specific queries only aim to obtain group membership information for specific multicast groups.

The router maintains a separate list of multicast group memberships for the subnetwork attached to each multicast-enabled interface. The list of multicast group members is updated as a result of incoming membership reports. If, after a group membership query, no membership reports for an active multicast group are received, the group membership is considered obsolete and removed from the list, meaning that no further multicast traffic is forwarded on the interface. In order to minimize the state required at the router, only active multicast groups for an interface need to be maintained at the router. Routers are therefore usually unaware of how many hosts on a given subnetwork are multicast group members.

The length of the periodic query interval may lead to delays in updating group membership details at the router. In order to speed up the join process, users can send unsolicited membership reports to inform a multicast-enabled router of a multicast group membership. Likewise, hosts can send a leave message for a particular group to a multicast-enabled router. Routers then send a group-specific query to determine if any other hosts are still interested in a specific multicast group prior to removing the entry from the membership list of a given interface.

Within a host, multicast group membership is specific to a particular network interface. The host application specifies which group membership to add or to drop on a particular network interface. Group membership reports are only sent by IGMP when a new multicast group membership is added for an interface. No reports are sent for successive joins to the same multicast group on the same interface. Likewise, a leave message is only sent for a multicast group when the last application drops the group membership on the interface.

An example of IGMP functionality is presented in Figure 2.5. The router periodically sends general membership queries, with the Destination Address (DA) being the all-systems multicast address 224.0.0.1 and the Group Address (GA) field in the IGMP message set to zero. After some time, an application joins the multicast group 239.0.10.10 via the socket interface. An unsolicited group membership report is sent to the multicast router, with both the DA and GA being set to 239.0.10.10, the address of the multicast group in question. The router adds the GA to the list of interested groups for the interface and passes the information to a multicast routing protocol. The router starts forwarding traffic for the multicast group on the interface. The router continues with its periodic queries during the multicast session, with the host replying with membership reports as long as interest exists. When the application leaves the multicast group, for example as a result of the user closing the application, a message is sent to inform the router that the host wishes to leave the group. The leave message is addressed to the all-router-multicast address 224.0.0.2 and provides the multicast group address in the GA

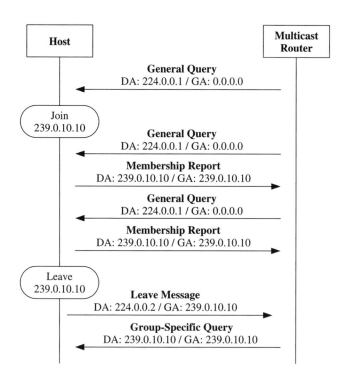

Figure 2.5   Example of IGMPv2 operation

field of the IGMP message. The router then sends a group-specific query to determine interest for the multicast group, prior to removing the entry from the list of the interface.

The latest version of the protocol, IGMPv3, introduces support for source filtering, which is required in order to support the SSM service model. Instead of subscribing to a multicast group, the receiver now subscribes to a multicast group as well as a list of sources associated with the group.

2.5.2   Multicast Listener Discovery

The MLD protocol, defined in (Karn *et al.*, 2004), is used by IPv6 routers to discover the presence of multicast listeners on its directly attached links as well as which multicast addresses are of interest to those neighbouring nodes. MLDv2 is the direct translation of IGMPv3, with IPv6 semantics.

There are two types of message in MLDv2, namely *multicast listener queries* and *multicast listener reports*. Multicast listener queries are sent by multicast routers to query the multicast listening state of neighbouring interfaces. A multicast router uses these queries to maintain and update its multicast address state for each of its attached links. The router does not need to maintain a detailed list of which hosts belong to which multicast group; it is sufficient for the router to know that there is at least one group member per network interface interested in the traffic of a multicast address. The multicast address state is a set of records, each containing a filter mode, source addresses and timers per multicast address.

There are three variants of query message defined in MLDv2. The *general query* message is sent periodically by multicast routers to find out which multicast addresses the systems on a link are interested in. If a multicast listener state changes (for example, if a multicast listener leaves a multicast group on a link or traffic from a source is not requested any more), multicast routers send specific queries to find out whether there are still systems interested in a particular multicast address or traffic from a particular source. For this, the *multicast-address-specific query* or the *multicast-address-and-source-specific query* are used.

IP nodes on a link keep an interface state for every IPv6 multicast address from which they want to receive traffic, per interface. The interface state contains a record including a filter mode and source list for each multicast address. When a node's interface state changes or a node receives a query, it sends an MLDv2 multicast listener report containing the appropriate multicast address records. The *current state record* returns the current state of the queried interface with respect to a single multicast address. The nodes send out the *filter mode change record* and the *source list change record* when local changes on the interface occur, the second one only when a source change does not affect the filter mode for a multicast address.

2.6   IP Multicast Routing

Multicast group management protocols are only concerned with gathering and maintaining membership information. Multicast routers need a way to communicate with each other to exchange information on group membership so that multicast traffic can be delivered to hosts on different subnetworks. This is done by means of routing protocols, which allow routers to establish paths over which multicast traffic can be sent. With multiple receivers, multicast packets need to be sent along several paths. The collection of paths over which multicast packets travel is referred to as the *multicast delivery tree*. The delivery tree should only consist of paths across subnetworks and routers that have multicast members. Only one copy of a packet must be sent on each path in the delivery tree. One copy of each packet is sent along each branch that leads to a group member. Multicast routing protocols carry out two tasks. They construct and maintain multicast delivery trees that connect multicast group members and forward multicast packets on the delivery tree.

Historically, the first experiments with IP multicast started at the beginning of the 1990s. The majority of routers at that time were not multicast-capable. The research community therefore set up an initially small network of interconnected workstations that served as multicast-enabled routers. The connectivity within this experimental network, called the MBone, was achieved with IP-encapsulated tunnels that crossed non-multicast-enabled routers. This virtual topology was flat, with research mostly focusing on mechanisms for non-hierarchical intradomain routing. Over time, however, the need to deploy mechanisms for hierarchical, interdomain routing became apparent (Almeroth, 2000). The following sections give an overview and describe the basic mechanisms of intra- and interdomain multicast routing protocols.

2.6.1   Multicast Distribution Trees

Multicast routing protocols use algorithms to create multicast distribution trees. Two different types of tree can be distinguished: *source-based trees* and *shared trees*.

A source-based tree has the source of multicast traffic as its root and forms a spanning tree through the network to all receivers. There are several algorithms used to create loop-free source-based trees. Reverse-Path Broadcast (RPB) creates a source-based tree optimized for each individual multicast source. Simply put, a multicast router verifies if a packet has arrived on the shortest path from its multicast source. If so, the packet is broadcast to all other interfaces, otherwise, the packet is dropped. The Truncated Reverse-Path Broadcasting (TRPB) algorithm optimizes the delivery such that routers with only one connection to the routing tree (said to be on a leaf of the routing tree) only deliver packets to those subnetworks with active group members. If this is not the case, the subnetwork is truncated and does not receive multicast traffic any further. The Reverse-Path Multicasting (RPM) algorithm optimizes the TRPB algorithm by allowing a complete branch of the delivery tree to be truncated, not only a leaf.

Unlike source trees that have their root at the source, shared trees use a single common root placed at some chosen point in the network, with this common root being shared by all sources of the group. With the root often called the Rendezvous Point (RP) or core, shared trees are also often referred to as RP or core-based trees. Sources forward all multicast traffic to the root, which then forwards the multicast traffic to all receivers in the network. There are two different types of shared tree, namely unidirectional and bidirectional shared trees. In a unidirectional tree, data can only flow downwards from the root to the branches of the tree. In a bidirectional shared tree, multicast traffic can flow up and down the tree in order to reach all receivers. In a unidirectional tree, sources need to find some other means to deliver multicast packets to the root. Shared trees have the advantage that only one tree needs to be constructed for each multicast group, hence multicast routers only need to keep a relatively small amount of state, which is proportional to the number of sources within the multicast group. Since the shared tree is common to all sources, the architecture of the tree may be not optimal for all sources. This may result in delivery delays of multicast traffic.

2.6.2 Intradomain Routing Protocols

Routing protocols are different in how they share information and compute routes. With link-state routing protocols, every node constructs a map of the connectivity of the network in the form of a graph showing which nodes are connected to which other nodes. With distance-vector algorithms, on the other hand, routers periodically share their knowledge of the entire network, but only with their neighbours. As neighbouring routers learn new information, they pass that information on to their neighbours and so on until, slowly but surely, the information makes its way across the entire network. Link-state protocols converge more quickly and are therefore more suitable for dynamically changing networks, but this comes at the cost of computational complexity.

While distance-vector and link-state routing protocols construct a source-based tree, shared tree protocols reuse the same tree for all sources within the group. Source-based tree protocols are also called dense-mode protocols, whereas shared tree protocols are known as sparse-mode protocols. All dense-mode protocols use a broadcast-and-prune mechanism and initially flood the network to find the optimal multicast route. This strategy is most suitable for areas with a dense concentration of group members. In contrast, sparse-mode protocols are more suitable for groups that are widely dispersed over a given network. Sparse-mode protocols usually offer better scalability than dense-mode protocols, since

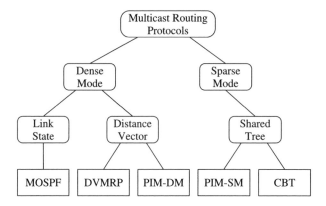

Figure 2.6 Intradomain routing protocols

only routers on paths from the source to all group members need to keep track of the
multicast group, while dense-mode protocols require state in all routers in the network.
Furthermore, sparse-mode protocols are more efficient since they do not flood the network
to establish the delivery tree. Rather, explicit join messages are used, so that traffic need
only flow across links that have been added. Sparse-mode protocols rely on a common
core or RP, which can easily turn into a single point of failure. Also, the delivery paths
in a shared tree are not always optimal compared with the source-based trees.

The intradomain routing protocols we consider are DVMRP, Multicast Open Shortest
Path First (MOSPF), PIM-DM, PIM-SM and Core-Based Trees (CBT). The two versions
of PIM, namely PIM-DM and PIM-SM, are used in environments in which members are
populated either densely or sparsely within the network. Figure 2.6 illustrates the classification of the most common intradomain protocols. A brief summary of these routing
protocols is provided in the following.

DVMRP Defined in (Waitzman, Partridge and Deering, 1988), DVMRP was the first
routing protocol to be deployed over the MBone. It is built on the principles of the
RIP unicast routing protocol and extends the concepts of RIP to multicast. DVRMP
uses reverse-path multicasting to generate the multicast delivery tree. It constructs a
separate routing table for multicast and uses the number of hops as its routing metric.
DVRMP supports the use of tunnels to bypass non-multicast routers and thus allows
for incremental deployment of IP multicast in the Internet. The major disadvantage of
DVRMP is that it does not scale well owing to its broadcast-and-prune behaviour and
the high amount of router state it requires.

MOSPF Defined in (Moy, 1994), MOSPF is the extension of the popular OSPF unicast
routing protocol that uses the link-state algorithm to compute the routes with minimum
protocol traffic overhead. Each router has a complete understanding over all links in
the network and uses that information to compute the routes to all destinations. Unlike
DVRMP and PIM-DM, the MOSPF does not flood the first packets of a multicast
transmission in order to build the routing tree. Group membership information is still
flooded to all neighbouring routers. A disadvantage of MOSFP is that it uses the routing
tables of OSPF for tree computation and therefore cannot operate with any other unicast

routing protocol. Furthermore, the computational complexity of the link-state algorithm increases proportionally with the number of multicast groups and group member sizes.

PIM-DM Defined in (Adams, Nicholas and Siadak, 2005), PIM-DM uses the underlying unicast routing protocol of the router for the construction of the multicast tree. Unlike MOSPF, however, PIM-DM is not dependent on any specific unicast routing protocol. The basic operation of PIM-DM is similar to that of DVMRP. It initially floods the first packets of a multicast session and then prunes back the leaves that have no multicast members. In contrast to DVRMP, PIM-DM does not keep the state of downstream routers for each multicast group, thus saving router resources. The downside of PIM-DM is that unnecessary packets may be forwarded to routers with no group members, requiring additional prune messages.

PIM-SM Defined in (Estrin *et al.*, 1998), PIM-SM is the most popular shared tree multicast routing protocol. While PIM-DM constructs a separate source-based tree for every source, PIM-SM uses a shared tree for all sources within a multicast group. The central element in PIM-SM is the RP, which is a preconfigured router somewhere in the network domain. Senders usually encapsulate their multicast packets in unicast data packets addressed to the RP. The RP decapsulates these packets and forwards them over the delivery tree to all group members. Receivers join by sending join messages towards the RP. The routers on the path towards the RP store status information for the group while passing the join request towards the RP, thereby constructing the delivery tree towards the new receiver. The discovery of the RP for a multicast address is done via a bootstrap protocol. A special feature of PIM-SM is that it is able to switch to a source-based tree when the data rate of a source exceeds a certain threshold. The intermediate routers will then send a join message towards the source and prune the tree towards the RP. This way, a more optimized delivery path can be constructed.

CBT Defined in (Ballardie, 1997), CBT is also a shared tree protocol, similar to PIM-SM. The root of the shared tree is called the core, which is comparable with the RP in PIM-SM. CBT has two major differences compared with PIM-DM. Firstly, CBT always uses a shared tree and is not able to construct a source-based tree. Secondly, the shared tree is bidirectional, whereas PIM-SM uses only a unidirectional shared tree. The complexity of constructing a bidirectional tree is slightly higher, but has a major advantage when packets sent by source towards the core cross branches of the multicast tree. Instead of first travelling all the way up to the core and then being sent down the tree to all members, intermediate routers directly deliver the packets to attached members while forwarding the packet towards the core. Thus, not all the packets have to cross the core to reach the receivers. CBT also uses unicast tables to obtain the next-hop router towards the core. As with PIM-DM, CBT is not dependent on a specific unicast routing protocol.

2.6.3 Interdomain Routing Protocols

The goal of interdomain multicast routing is to provide a hierarchical and scalable Internet-wide multicast service. Currently, two solutions exist. The first solution relies on the Multicast Border Gateway Protocol (MBGP), a straightforward extension to the Border Gateway Protocol (BGP) used in unicast routing. MBGP is used to advertise multicast routes between domains. In addition to MBGP, PIM-SM is used for the construction of delivery trees between domains, whereas the Multicast Source Discovery Protocol (MSDP) is used for the announcement of sources between domains. MBGP in combination with PIM-SM and MSDP is seen as a near-term solution for interdomain multicast routing, as it currently works well but does not scale to large multicast group populations.

The second solution relies on the Border Gateway Multicast Protocol (BGMP). BGMP avoids interdomain dependencies by relying on a strict address allocation scheme.

The scalability issues that result from MBGP used in combination with PIM-SM and MSDP can be traced to MSDP. MSDP has a relatively high message overhead that does not scale well to large numbers of sources in the network. Also, dynamic group changes, caused either by sources of bursty traffic or by frequently joining or leaving group members, may not be handled in a timely manner owing to latencies in announcing sources and constructing the multicast tree. An overview of MBGP, MSDP and BGMP is given below.

MBGP Based on the well-known protocol BGP used for unicast interdomain routing, MBGP offers substantial abstractions and control among domains. With BGP, each domain is able to employ its own routing protocol, while the routing policies of BGP determine the best link to external networks. MBPG extends BGP with the ability to exchange information about multicast routes between the domains. Without MBGP, every root would need to know the entire flat multicast topology. With MBGP, the required knowledge is limited to only the topology in the local domain and the paths to each other domain. MBGP enables the discovery of the next hop to a host or a router in an external network, although it does not provide means to construct the routing tree towards this host. As a result, PIM/SM can be used in order to construct a multicast tree between different domains that have multicast group members.

MSDP MBGP assumes that the multicast routing protocol will have one RP in every domain for a particular multicast group. The goal in the long term is to have a single multicast RP within the entire Internet, rather than having multiple RPs in every domain. Therefore, the basic problem is how to inform an RP in one domain about sources present in other domains. MSDP, defined in (Fenner and Meyer, 2003), attempts to solve the problem by announcing the existence of active sources to other domains. MSDP runs on the same router that acts as the RP of the domain. When a source registers with the delivery tree, MSDP detects the new active source and floods information on the source to all MSDP peers in the RPs of neighbouring domains. The flooding algorithm includes reverse-path-first checks to prevent message looping. The peer RP will check if it has any member state for a given group. If so, it will join the PIM-SM towards the source address advertised by the MSDP announcement.

BGMP Defined in (Thaler, 2004), BGMP is designed to interoperate with any intradomain multicast routing protocol. It requires each multicast group to be associated with a single root/core and constructs a shared tree of domains. More specifically, BGMP associates a range of multicast addresses with an administrative domain. An initiator of a multicast session will thus use a multicast address if its domain becomes the root domain for the possibly Internet-wide multicast tree. The root domain assignment is based on the assumption that the group initiator will contribute as a significant source to that group. BGMP runs on domain border routers and constructs a bidirectional shared tree that connects individual multicast trees built within a domain. Hence, these border routers also run the protocols used for the interdomain routing. For multicast tree construction, BGMP uses routes advertised by BGP. BGMP, however, requires a strict address allocation scheme. The Multicast Address Set Claim (MASC) protocol can be used for interdomain multicast address set allocation. Nonetheless, more sophisticated address allocation schemes are still seen to be required (Almeroth, 2000).

2.7 Reliable Delivery of Multicast Traffic

Delivery requirements usually differ between applications. For many applications, reliable delivery is not essential. Multimedia streaming applications, for example, are able to live with occasional packet losses. The Real-Time Protocol (RTP) is widely used as a transport protocol for applications that require low delivery latencies while being insensitive to packet losses. Applications that require delivery guarantees but are not sensitive to delivery latencies typically make use of TCP. Although TCP is the most commonly used reliable transport protocol today, it can only be used for unicast delivery.

Multicast applications that require reliable delivery have two options. They can either rely on specialized transport protocols operating on top of UDP or they can bypass UDP completely and make use of reliable transport protocols operating directly above the IP layer. This is illustrated in Figure 2.7.

Designing a suitable reliable multicast protocol poses several challenges. A reliable multicast protocol must be able to operate in a scalable manner. A large number of receivers can easily result in the sender being overwhelmed by the amount of back traffic generated by the positive or negative acknowledgements sent by receivers. This is known as the feedback implosion problem. Also, if multicast group members are globally distributed across different heterogenous network segments, packet losses are likely to be uncorrelated. A large number of uncorrelated packet losses translates into a large number of retransmissions, which in turn can cause network congestions.

2.7.1 Early Approaches for Reliable Multicast

Several proposals for reliable multicast attempt to achieve protocol scalability by reducing the number of ACKs or NACKs that are sent as feedback to the source. The Reliable

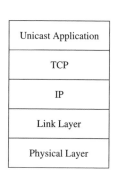
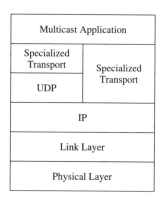

Figure 2.7 Protocol stacks for reliable transport

Multicast Transport Protocol (RMTP), defined in (Lin and Paul, 1996), introduces designated receivers that act as hierarchically organized repeaters and aggregators of acknowl-
egement feedback in the network. Instead of directly sending control messages to the
multicast source, receivers send control traffic via unicast to designated receivers in their
local proximity. A designated receiver can provide local recovery, thus keeping necessary retransmissions local. In addition, designated receivers aggregate control traffic and
request missing packets only from the next designated receiver in the hierarchy.

Although RMTP is able to alleviate the feedback implosion problem, it still puts an
additional burden on a small number of designated receivers that need to perform the
required retransmissions. Scalable Reliable Multicast (SRM), defined in (Floyd *et al.*,
1997), allows any receiver to provide repair. SRM makes the receivers responsible for
requesting retransmissions from neighbouring receivers. Receivers send status information
via low-frequency session messages to the multicast group. This allows other receivers
to learn about their neighbours and the status of their neighbours. Receivers are required
to cache received data so that every receiver can respond to a repair request. A back-off
mechanism is used to reduce the traffic for repair requests and the resulting retransmissions.

Pretty Good Multicast (PGM), defined in (Speakman *et al.*, 2001), does not rely on
receivers to perform repair, but instead requires routers in the network to aggregate control
traffic. PGM is an example of a protocol that sits on top of the IP layer and bypasses UDP.
PGM is designed to provide reliable delivery for real-time applications. Availability of
repairs is determined by a sliding retransmission window, and repairs within this window
are performed either by the source or by designated local retransmitters. Receivers request
retransmission via NACKs, which are sent to their local PGM-aware router. NACKs are
suppressed locally and sent back via unicast following the reverse multicast distribution
tree. Routers along the path filter duplicate NACK requests. Although scalable, PGM
requires an upgrade of router infrastructure.

The Multicast File Transfer Protocol (MFTP), defined in (Miller *et al.*, 1997), is a
scalable approach for reliable delivery of files from a sender to multiple receivers. MFTP
breaks the file to be transferred into large blocks of thousands or tens of thousand of
packets and performs retransmissions in response to the NACKs from the receiver population. Unlike with RMTP, where receivers send NACKs for each lost or erroneously
received packet, a receiver in MFTP sends a single NACK for a complete block. The
NACK takes the form of a bitmap, where each bit indicates whether or not a packet in
the block has been received successfully. The sender determines which packets need to
be retransmitted by OR-ing the NACK bitmap from all receivers for a given block. At the
receiver, NACKs are usually spread by a back-off timer, thus avoiding a sudden implosion of NACKs at the sender. NACKs can also be aggregated by intermediate routers
in order to provide higher scalability. In MFTP, the sender does not send the requested
retransmission after each block, but instead awaits the completion of transmission for all
blocks of the file.

2.7.2 *Recent Developments in Reliable Multicast*

Applications vary in terms of their requirements with respect to ordering and delivery guarantees, receiver scalability, real-time feedback, support for underlying network
topologies, knowledge of group membership at the source and so forth. Whereas initially

the focus of the research community was on creating a standard for reliable multicast, the diverse requirements of different applications make it nearly impossible to provide a one-size-fits-all solution to reliable multicast.

The IETF has therefore defined a framework for the standardization of specialized reliable multicast transport protocols that relies on protocol *building blocks*. A building block is defined as a logical protocol component that is able to provide its services via a well-defined API to other protocol building blocks or a protocol client. The most important protocol building blocks are as follows:

NACK-based reliability This protocol building block defines the detection, notification and recovery of data loss based on NACK feedback. The protocol building block must address how lost packets are specified within NACK messages and how NACK implosion can be prevented.

FEC coding This protocol building block provides packet-level FEC functionality, which is used to reduce the number of packet losses. The building block specifies what FEC coders and decoders are used and how FEC packet naming is performed.

Congestion control This protocol building block specifies mechanisms to avoid congestion in the network and to allow multicast flows to behave fairly with respect to other traffic in the network. Two main approaches are considered: a single-rate congestion control approach regulated by the source and a receiver-driven multirate approach that allows recipients to receive traffic at different rates according to network congestion levels.

Generic router support This protocol building block defines how functional components can take advantage of support in the routing infrastructures. It specifies a signalling protocol between routers and defines algorithms that allow the router to perform the support function.

Tree configuration The tree configuration building block provides mechanisms for constructing and managing a logical tree in order to interconnect agents that assist the source in the delivery of data, for example by performing retransmission or aggregating feedback.

Other protocol building blocks include data security, common protocol headers and protocol cores. Details on these building blocks can be found in (Whetten *et al.*, 2001). Two examples of multicast reliability protocols that rely on the protocol building blocks defined in (Whetten *et al.*, 2001) are the File Delivery Over Unidirectional Transport (FLUTE) protocol and its key building block, Asynchronous Layered Coding (ALC). Both are described below.

2.7.3 Asynchronous Layered Coding

The Asynchronous Layered Coding (ALC) protocol, defined in (Luby *et al.*, 2002a), is a massively scalable reliable content delivery protocol. ALC adds session management functionality, congestion control and reliability to IP multicast, without sacrificing any of its scalability properties. ALC requires no feedback packets between the sender and all receivers. This makes it suitable for unidirectional links and allows senders to support an increasing number of receivers without impacting their load or outgoing rate. Receivers, on the other hand, have the perception of participating in a dedicated session, with the reception rate being adjusted to the congestion levels experienced along the path from the sender to the receiver.

ALC can support a large number of receivers, potentially in their millions, and is able to deliver arbitrarily large content objects such as files. ALC allows each receiver to initiate a delivery session for an object asynchronously and facilitates the reception of data at a rate that corresponds to the maximum fair and allowable bandwidth between the sender and the receiver. ALC makes use of several of the above-mentioned protocol building blocks such as the multirate congestion control building block to provide feedback-free congestion control, the Layered Coding Transport (LCT) building block (Luby *et al.*, 2002b) for in-band session management functionality and the FEC building block (Watson Luby and Vicisano, 2007) for reliability.

An ALC session is identified by a sender address and a unique Transport Session Identifier (TSI). Each session typically comprises multiple channels that originate at a single sender. These channels are used for some period of time to carry packets corresponding to the transmission of one or more objects in which the receivers are interested. If multiple objects are sent in a session, a Transmission Object Identifier (TOI) is used to distinguish packets belonging to different objects. Congestion control is usually applied over all packet flows belonging to a single session. When using a multirate congestion control protocol, data from the sender is usually encoded in multiple layers, which are mapped to different channels. Each channel is typically identified by the combination of the address of the sender and that of the group. These channels may or may not be provided at different rates. Receivers can adapt their reception rate by joining and leaving channels in order to receive data according to their available network bandwidth.

Before sending the objects within a session, the ALC protocol applies mechanisms provided by the FEC building block to generate encoded symbols for the object to be transmitted. Each object is individually encoded, and the resulting symbols are fed to different channels of the session. Receivers that have subscribed to one or more channels simply wait until they have received a sufficient number of packets to allow them reliably to reconstruct an object. As a result, there is no need for receivers to send retransmission requests.

In order for ALC to work, a sender needs to provide a session description to the receiver. This session description contains all information necessary for a receiver to initiate the reception of an object. The session description is obtained by means of out-of-band signalling. At a minimum, the following information must be provided:

- the IP address of the sender;
- the TSI to be used for the session;
- an indication if the session carries packets for more than one object;
- the multirate congestion control building block utilized;
- the number of channels in the session;
- the address and port number of each channel in the session;
- information on the header extension to be used;
- information on the packet authentication.

The ALC packet format is shown in Figure 2.8. ALC operates on top of UDP and adds an LCT header and an FEC payload ID to the UDP header. The payload of the ALC

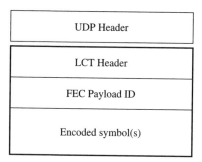

Figure 2.8 ALC packet format

packet is made up of the encoded symbols for the transmitted object. ALC packets are not dependent on a particular IP version and can thus be sent over IPv4 or IPv6. Although ALC is specifically designed for protocols using IP multicast, it can also be used with unicast protocols.

2.7.4 The FLUTE Protocol

FLUTE is a protocol for unidirectional delivery of files over IP-based communication networks (Paila *et al.*, 2004). FLUTE is intended for the delivery of small and large files to many receivers, using delivery sessions that can last several seconds or longer. FLUTE builds on the ALC protocol building block and is hence massively scalable. FLUTE extends ALC with the following features:

- It defines a file delivery session, which is built on top of ALC but also includes transport details and timing constraints.
- It provides in-band signalling of transport parameters for the ALC session as well as properties of transmitted files.
- It defines mechanisms for multiplexing multiple files within a single session.

Figure 2.9 shows the internal building block structure of FLUTE. Not all of ALC's internal building blocks are mandatory in FLUTE. As an example, properly provisioned and controlled communication networks such as mobile communication and satellite systems do not require the congestion control building block. Likewise, the use of FEC is optional.

In FLUTE, the concept of an ALC session is referred to as a file delivery session, with an ALC object representing the file to be transferred. By default, all ALC packets in a FLUTE session must have a TOI field. FLUTE introduces the File Delivery Table (FDT) as a means of providing an in-band description of various relevant file attributes. Logically, an FTD is a set of file description entries. Each file description entry must provide the TOI for the file and the URI identifying the file. Within a file delivery, an FTD is usually delivered as an FTD instance. An FTD instance contains one or more file description entries for the FTD. Instances of the FTD are assigned a TOI field value of zero, while values for a TOI greater than zero are used for the files of the delivery session. FTD instances are described in an Extensible Mark-up Language (XML) document that

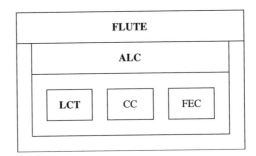

Figure 2.9   Internal building blocks of FLUTE

follows a predefined structure. The description of a file delivery session follows the recommendations provided by ALC. A receiver needs to know the IP address of the sender, the TSI of the session, the number of channels in the session, the destination IP address, the port number of each channel in the session and an indication of whether the session is a FLUTE session.

Figure 2.10 shows a sequence of file delivery steps at the receiver. The receiver is informed of the IP address of the sender, the TSI to be used for the session, available destination addresses and port numbers. This information is obtained via file delivery session descriptions. The receiver then joins one or more channels. The receiver verifies if the received ALC packets are associated with the desired TSI. It then demultiplexes incoming packets according to their TOI and stores these packets. The receiver is able to recover an object when a sufficient number of packets containing encoded symbols for the transmission object have been received. The number of required packets for successful decoding depends on the FEC coding scheme. Received objects with a TOI value of zero are assumed to be FDT instances. For these objects, the receiver populates its FTD database with the appropriate file descriptions. For objects other than FTD instances, the receiver queries the FTD database and, if file properties are available, assigns these to their corresponding files. A check whether the received content matches the corresponding descriptions in the FDT database is also performed.

2.8   Multicast Flow and Congestion Control

In this section, we briefly touch upon the basic mechanisms for multicast flow and congestion control. Multicast flow control can be seen as a set of mechanisms that aims at matching the data transmission rate of the source to the available capacities at the receivers as well as the available bandwidth on route to the receivers. The primary aim of multicast flow control is thus to prevent receivers being overwhelmed by the rate of data transfer at the sender. Multicast congestion control, on the other hand, is responsible for adapting the rate of data transfer in response to network conditions such as congestions and also according to the principle of fairly sharing available resources among multiple data flows. It thus addresses a more global issue, as it deals with fairness not only among members of a multicast session but also between different multicast sessions and any other traffic present on a given link. Multicast flow and congestion control are strongly interrelated and therefore often performed jointly.

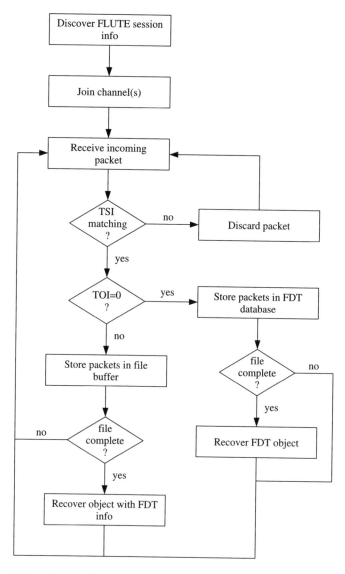

Figure 2.10 High-level sequence diagram of file delivery at a FLUTE receiver

Most existing multicast applications operate on top of UDP, which only provides best-effort delivery. Unlike TCP, UDP does not provide any back-off-based congestion control. On congested links with limited network resources, UDP-based flows can starve out TCP-based flows, forcing TCP-based senders to overthrottle to a very low transmission rate. One of the objectives of congestion control is thus to achieve intersession fairness, making multicast flows ideally behave similarly to TCP or TCP-like traffic. One obvious way to avoid congestion in the network is to overprovision network resources or to apply a strict regime of resource reservation. Such controlled or provisioned environments, of which mobile networks is one example, do not require congestion control mechanisms.

On the global Internet, however, suitable multicast congestion control mechanisms are required. Another issue that flow and congestion control must address results from the heterogeneity of potentially globally distributed multicast receivers and their intermediate network segments along the path. It is often desirable to have the transmission rate of the receiver adapt to the available bandwidth of the path to each receiver. Limiting the transmission rate to the bottleneck link capacity of the worst-case receiver would penalize other session members. This problem is also referred to as interreceiver fairness.

Flow and congestion control schemes can be grouped into single-rate or multirate approaches. With a single-rate approach, multicast session data is sent at a single rate to all receivers. Although this approach may achieve intersession fairness, single-rate schemes cannot guarantee interreceiver fairness in the presence of heterogeneous receivers. Therefore, multirate schemes are usually preferred, especially for large receiver groups. Multirate schemes for flow and congestion control transmit multicast session data at multiple rates. Multirate transmission can be achieved in different ways. This can be achieved by encoding the original data into a number of data streams and then simulcasting or replicating these streams. Each receiver can select a stream at a rate that is appropriate with respect to the available capacity at the receiver. This approach adds additional overhead, as redundant information is sent in parallel streams. A more efficient scheme for multirate transmission is layered multicast, where data is divided into several layers. Each layer is transmitted to a different IP multicast group. Layers can be coded at the same or at different rates. In cumulative layering schemes, all receivers have to subscribe to a base layer that contains all essential features to receive a data stream correctly. Based on bandwidth availability, receivers can also join more layers in order to receive data at a higher rate. Depending on the type of data, this may result in higher reception quality for video streaming application or faster download times for file transfers. Non-cumulative schemes do not require hierarchical layering, and receivers can join any of the available layers. Multirate congestion control can also be achieved by transcoding multiple flows in intermediate network nodes or by filtering flows in the routers along the multicast delivery path. Please refer to (Li and Zhang, 2002) for a thorough discussion of flow and congestion control protocols for IP multicast.

2.9   Multicast in Mobile and Wireless Networks

All multicast protocols described up to this point assume static connectivity between nodes. The number of mobile nodes is expected to increase throughout the Internet in coming years. Existing multicast protocols may require modifications in order to operate efficiently when hosts are mobile.

What is the impact of host mobility on IP multicast protocols? In contrast to wired links, which are in most cases symmetrical and usually provide a fair amount of bandwidth, wireless links are mostly asymmetrical and often have bandwidth limitations. The asymmetry may be inherent to characteristics of the link, as is the case with unidirectional broadcast networks, or may be the result of varying network conditions in the uplink and downlink of a cellular network. Wireless channels typically have a higher error rate, with packets errors occurring more frequently as a result. Also, the mobility of hosts may necessitate dynamic changes to the routing structure of the multicast tree. In some cases, the new point of network attachment of a mobile host may not provide the necessary infrastructure to support IP multicast or provide the required bandwidth for the multicast session in question.

Mobile IP is the standard protocol for host mobility on the Internet (Perkins, 2002). Mobile IP for IPv6 or MIPv6, defined in (Perkins, 2002), will most likely become a fixed component of IPv6. Bidirectional tunnelling and remote subscription are two ways to support data delivery to mobile receivers in Mobile IP. How both approaches can be used to support IP multicast is described in more detail below.

2.9.1 Bidirectional Tunnelling

Bidirectional tunnelling is proposed in (Perkins, 2002) and assumes that a Mobile Host (MH) sends and receives data for a particular multicast group via its Home Agent (HA). While the MH is roaming and attached to a foreign network, it sends and receives multicast traffic over a bidirectional tunnel that is constructed using IP encapsulation to its HA. This is done independently of the location of the MH. Group membership messages are also exchanged with the home network via the bidirectional tunnel. This approach, however, requires the HA to be a multicast-enabled router. Alternatively, the HA can proxy group membership messages to a multicast router on its home network.

The main advantage of this approach is that it does not require the multicast tree to be reconstructed whenever a host involved in a multicast session changes its point of attachment. Therefore, this approach is suitable for hosts that are highly mobile. Also, this approach is simple to implement. On the other hand, the HA can quickly become a performance bottleneck with large multicast groups. Also, traffic is routed in a suboptimal fashion, especially when the host is located far away from its home network or the source of the multicast group. Bidirectional tunnelling is not efficient, as multicast traffic is transmitted over multiple tunnels to the receivers in the foreign network. This is known as the multicast avalanche problem.

2.9.2 Remote Subscription

With remote subscription, the MH joins the multicast group via a local multicast router in its foreign network. Membership handling is carried out using IGMP or MLD. The local multicast router that receives a membership request simply joins the delivery tree of the group in the usual fashion.

The advantages of remote subscription is that it is efficient in terms of processing at routers and bandwidth consumption. Multiple group members on the same foreign network receive multicast traffic on an optimal path from the local multicast router. The approach is also independent of the underlying mobility protocol.

Frequent joins and leave messages lead to a higher reconstruction overhead on the multicast delivery tree, especially when a large amount of group members is mobile. Also, the delay for tree reconstruction may result in packet losses within the session. Frequent tree reconstruction is especially costly with source-based trees, and therefore shared trees are preferable for data delivery with a fixed root. SSM cannot be supported with remote subscription. The MH must change its Care of Address (CoA) when attaching to another network. The MH will not receive any multicast traffic until it resubscribes to the new CoA.

2.9.3 Extensions for Mobile Multicast

Several extensions to bidirectional tunnelling and remote subscription have been proposed to overcome some of the inefficiencies of these approaches. The most relevant of these extensions are briefly described in the following.

Preserving the Multicast Nature of Traffic

The main issue with bidirectional tunnelling is that it does not preserve the multicast nature of traffic. In order to solve the multicast avalanche problem, Lee (2001) proposes the use of Explicit Multicast (XCast), which is defined in (Boivie *et al.*, 2007). Instead of sending multicast data to the same session on separate tunnels, the HA performs multicast-in-XCast encapsulation. The HA needs to keep track of the individual receiver addresses for each multicast group. The HA encodes the list of addresses in an XCast header. Only a single packet is forwarded by the HA, with the XCast-enabled router replicating the packet where needed. The number of receivers that XCast can support is very limited. If the list of receivers is too long, the HA generates several XCast headers and sends multiple copies of the same data with each of the generated XCast headers.

Seamless Session Handover

Receiving multicast traffic via remote subscription has several advantages with respect to processing at routers and bandwidth consumption. The subscription process with the local multicast-enabled router may be time consuming and thus cause session data to be lost. A combination of bidirectional tunnelling and remote subscription is proposed in (Jelger and Noel, 2002) in order to make session handovers more seamless. The MH always uses remote subscription and joins the session at a local multicast-enabled router. At the same time, the MH is subscribed to the session via bidirectional tunnelling but does not receive any traffic during normal operation. When the MH is about to subscribe on a foreign network, it activates the tunnelling from the home network via a membership report. The HA starts tunnelling the data to the MH, while the remote subscription is in process, thus avoiding a loss of data. After the subscription process finishes, the MH can send a *hold* message to the HA to stop the redundant forwarding of packets via the bidirectional tunnel. The hold message includes the list of multicast group addresses to which the mobile user is subscribed. The HA remains subscribed to the multicast sessions.

2.9.4 Agent-Based Multicast

A Multicast Agent (MA) can be used to provide mobility to multicast receivers. The MA is located in the foreign network and joins the multicast delivery tree on behalf of the user. After joining the multicast tree, the MA is responsible for forwarding (or tunnelling) multicast traffic to the multicast host, while at the same time hiding the mobility of the MH from the main distribution tree. Generally speaking, agent-based approaches focus on improving the operation of multicast using the FA in Mobile IPv4. Mobile Multicast Protocol (MoM), defined in (Harrison *et al.*, 1997), aims to solve the tunnel convergence problem. For the same multicast group, an FA can receive multiple Mobile IP tunnels from different HA. In MoM, the FA selects only one HA as a Designated Multicast Service

Provider (DMSP), which tunnels the multicast for all HAs to the FA. The FA forwards the packets on the foreign network using native IP multicast. Range-Based Mobile Multicast (RBMoM), defined in (Lin and Wang, 2000), tries to strike a balance between the shortest delivery path and the tree reconstruction overhead. Similar to MoM, it also uses a DMSP to solve the tunnel convergence problem at the FA. However, instead of having a fixed HA, RBMoM introduces a Multicast HA (MHA), which joins the multicast tree on behalf of the users and tunnels the traffic towards the HA. Each MHA has a certain service range; thus, if an MH changes its point of attachment to another network outside the range of its old MHA, an MHA that is topologically closer joins the multicast delivery tree and starts forwarding multicast traffic to the FA. The performance of this approach is dependent on the service range. The service range is a value that can be adapted to group sizes and mobility models.

Finally, Multicast by Multicast Agents (MMAs), defined in (Suh, Shin and Hee, 2001) operates by having an MA forward data onto a foreign network using native IP multicast, similar to RBMoM. However, an MA can also take the role of a Multicast Forwarder (MF), in case the MH moves to another network. An MF forwards the multicast data for a multicast group to the MA on the new network. The range of the MF is, however, not limited. In case an MH that is currently receiving traffic from a multicast group over an MF moves to a new network, with an MA that is already served with the same session data by another MF, an MF selection occurs and one MF will stop forwarding multicast data.

2.10   Summary

There are two service models for IP multicast. The ASM service model is the most widely used service model today. Its lack of access control makes it vulnerable to security threats, and interdomain routing issues make global deployment problematic. The SSM service model addresses these problems, but requires the receiver to know the multicast group address as well as the source address of the sender. IP multicast addresses have a dedicated address space in both IPv4 and IPv6. Multicast group management allows multicast-enabled routers on a subnetwork to discover whether local hosts wish to receive traffic from a particular multicast group. Multicast group management protocols thus only run between the local hosts and routers on the subnetwork. In IPv4 networks, IGMPv2 is used for the collecting group membership for the ASM service model, while IGMPv3 is used for implementing the SSM model. In IPv6, group management for the ASM service model is handled by MLDv1, whereas MLDv2 handles group management for the SSM service model. Multicast routers use group membership information to establish efficient delivery trees from the multicast source or RP towards the receiver population.

As most IP multicast applications operate on top of the UDP, the delivery of IP multicast packets is inherently unreliable. Reliable multicast protocols apply error control mechanisms such as coding and packet retransmissions to recover from packet errors. Mechanisms that require receivers to provide feedback to the source do not scale well and can lead to a feedback implosion at the sender. Error coding approaches can be used to increase the scalability of reliable multicast protocols. The IETF has developed a framework for the design of reliable multicast protocols that relies on well-defined protocol building blocks. The FLUTE protocol, used for the scalable and reliable multicast delivery of data files, is designed based on the ALC protocol building block. ALC in turn makes use of several other protocol building blocks such as the multirate congestion control building block and the LCT building block for in-band session management functionality. FLUTE is used in third-generation mobile networks to provide reliable delivery of multicast traffic.

The mobility of multicast group members poses several challenges for IP multicast. The goal is to allow multicast sessions to continue without disruption when either a source or a group member changes its point of network attachment. Ideally, multicast host mobility should be achieved without major impact on the underlying topology of the multicast delivery tree. Mobile IP provides mechanisms such as bidirectional tunnelling and remote subscriptions for dealing with host mobility. Both bidirectional tunnelling and remote subscription are not always optimal in terms of bandwidth consumption and cost of tree reconstruction. Several extensions to Mobile IP, such as agent-based multicast, attempt to overcome some of these inefficencies.

3

An Overview of Third-Generation Networks

3.1 Introduction

The purpose of this chapter is to provide an overview of the most important architectural and procedural aspects of UMTS and CDMA2000 networks.

For UMTS, we first describe the WCDMA air interface. In mobile networks, the air interface is the bottleneck with respect to system capacity. For the WCDMA air interface, we describe the spreading and despreading operation, multipath reception by means of a Rake receiver, power control and soft handover. We then introduce the radio interface protocol architecture. Finally, the channel mappings between the physical, transport and logical channels are depicted. Similarly, for CDMA2000, we first introduce the different CDMA2000 air interface standards, namely CDMA2000 1xRTT, CDMA2000 EV-DO and CDMA2000 EV-DV. We then describe the architecture of the radio access network as well as the the radio interface protocol architecture for CDMA2000. We then briefly introduce the EV-DO Revision 0 standard, paying particular attention to its channel structure. Finally, we describe the architecture of the CDMA2000 core network, alongside its associated procedures for mobility management.

3.2 Radio Access and Networking in UMTS

A UMTS network consists of two land-based network segments: the UMTS Terrestrial Radio Access Network (UTRAN) and the Core Network (CN). The CN itself is further divided into the circuit- and packet-switched domains. The CN in the packet-switched domain consists of two logical network nodes, namely the Gateway GPRS Support Node (GGSN) and the Serving GPRS Support Node (SGSN). The UTRAN consists of the Radio Network Controller (RNC), which connects to multiple base stations. In UMTS terminology, base stations are referred to as Node Bs. A single Node B may serve one or more cells. The RNC connects to the CN via the SGSN.

Multicast in Third-Generation Mobile Networks   Robert Rümmler, Alexander Gluhak and A. Hamid Aghvami
© 2009 John Wiley & Sons, Ltd

3.2.1   Air Interface

WCDMA is the air interface used in UMTS networks. WCDMA supports two basic modes of operation, FDD and TDD modes. In the FDD mode, separate 5 MHz carrier frequencies are used for the uplink and downlink respectively, whereas in TDD mode only one 5 MHz carrier is time-shared between the uplink and downlink. WCDMA in FDD mode is the more dominant and widespread air interface, with the TDD mode based heavily on the FDD mode. We only consider WCDMA in FDD mode. Details on UMTS TDD mode can be found in (Holma and Toskala, 2007).

Spreading and Despreading

In WCDMA, user information bits are spread over a wide bandwidth by multiplying the user data with quasi-random bits called chips. The chip rate of 3.84 Mcps in WCDMA leads to a carrier bandwith of approximately 5 MHz. The spreading operation is the multiplication of each user data bit with a sequence of chips. The resulting spread data is at a rate of the chip rate or spreading factor times the user data rate. The wideband signal is then transmitted across the wireless channel to the receiving end. During despreading, the transmitted signal is multiplied with the same code chips that were used during the spreading operation. This allows the original user bit sequence to be recovered perfectly, provided perfect synchronization between the spread user signal and the spreading code can be achieved. The increase in signalling rate by the spreading factor corresponds to a widening of the occupied spectrum of the spread user data signal. Despreading restores the original bandwidth, which is proportional to the data rate of the signal.

In WCDMA, the processing gain resulting from the spreading and despreading operation together with the wideband nature allows the same frequency to be reused in adjacent cells. This results in high spectral efficiency. Having many users share the same wideband carrier provides interferer diversity, resulting from the multiple-access interference from many users averaging out. Also, with a wideband signal, the different propagation paths of the radio signal can be resolved at higher accuracy than with signals at a lower bandwidth. This results in higher diversity against fading. However, tight power control and soft handover are required to avoid one user's signal blocking the signals of other users (Holma and Toskala, 2007).

Multipath Radio Channels and Rake Reception

Radio propagation in the land mobile channel is characterized by multiple reflections, diffractions and attenuation of signal energy. These are caused by natural obstacles such as buildings, hills and so on, resulting in what is known as multipath propagation. The signal energy may arrive at the receiver across clearly distinguishable time instants. If the time difference of the multipath components is larger than the duration of a single chip, the WCDMA receiver can separate those multipath components and combine them coherently to obtain multipath diversity. For certain time delay positions, there are usually many paths nearly equal in length along which the radio signal travels. As a result of this, signal cancellation, called fast fading, takes place as the receiver moves across even short distances.

These fading dips make error-free reception of data bits very difficult, and counter-measures are needed for WCDMA. The delay dispersive energy is combined by utilizing multiple correlation receivers or Rake fingers allocated to the individual delay positions on which significant energy arrives. Fast power control and the inherent diversity reception of the Rake receiver are used to mitigate the problem of fading signal power. Strong coding and interleaving as well as retransmission protocols are used to add redundancy and time diversity to the signal and thus help the receiver in recovering the user bits across fades. CDMA signal reception can be performed on the basis of the following operating principles (Holma and Toskala, 2007):

1. Identify the time delay positions at which significant energy arrives, and allocate correlation receivers or Rake fingers to those peaks.
2. Within each correlation receiver, track the fast-changing phase and amplitude values originating from the fast-fading process and remove them.
3. Combine the demodulated and phase-adjusted symbols across all active fingers and present them to the decoder for further processing.

In typical implementations of the Rake receiver, processing at the chip level is done in Application-Specific Integrated Circuits (ASICs), whereas symbol-level processing is implemented in Digital Signal Processing (DSP). Further details on Rake reception can be found in Holma and Toskala (2007).

Power Control

As power is a resource that is shared between all users, tight and fast transmit power control is a very important aspect of CDMA systems. Consider a receiver and two transmitters, with one transmitter located closer to the receiver, the other transmitter further away. If both transmitters transmit simultaneously and at equal power, then the receiver will receive more power from the transmitter that is closer. This makes it difficult for the receiver to resolve the signal of the transmitter that is further away. Having a transmitter close to the receiver that is transmitting at high power effectively jams the communication channel for other transmitters that are further away. This is referred to as the near-far problem in CDMA systems and must be combatted by means of efficient transmit power control, especially on the uplink.

A rough means to control the uplink transmit power is open-loop power control. The terminal estimates the attenuation on the radio channel by listening to a pilot or beacon signal sent by the base station at a known power level and regulates its transmit power according to this estimate. The problem in an FDD air interface such as WCDMA is that, owing to frequency separation between the links, the uplink and downlink fast-fading processes are more or less independent. This method is therefore not very accurate.

More accuracy can be attained with closed-loop power control, where the base station measures received Signal-to-Interference Ratio (SIR) levels, compares them with a target level and, based on the outcome of this comparison, orders the mobile terminal either to increase or to decrease the transmit power level. In WCDMA, this happens 1500 times per second, and hence the power control rate is 1.5 kHz. This is fast enough to track pathloss and shadowing, and even fast fading of mobiles at low to moderate speeds. Closed-loop power control can be decomposed into outer-loop and inner-loop components. Outer-loop power control is a slow activity consisting of adjustments in the target SIR level, based on the quality requirements and the current propagation conditions. Inner-loop power control is the fast ordering and adjustment process carried out once per slot to meet the target SIR.

On the downlink, since there is only a single source in a cell, power control is not needed to overcome the near-far problem. Instead it is used to compensate for fading deeps and to aid mobiles at the cell edge suffering from increased intercell interference.

Soft Handover

Compared with GSM, UMTS supports two new types of handover, namely soft and softer handover. In both cases, communication between a mobile terminal and the network takes place over two (or more) air interface channels concurrently. With softer handover, the two channels are associated with two different sectors served by the same Node B. During soft handover, the mobile terminal is connected to the network via multiple Node Bs, which may even be controlled by different RNCs. Being connected to more than one base station during soft handover provides a diversity advantage. If the same signal is transmitted via different propagation paths that exhibit no or little correlation, then the probability that at least one of the paths delivers sufficient signal quality at the receiver at any given time is higher than when only a single path is relied upon. This effect is referred to as macrodiversity.

In the uplink direction, a single signal transmitted by a mobile terminal is received by multiple Node Bs during soft handover. These base stations constitute the so-called active set. The different received signal copies are combined by the Serving RNC (SRNC). If Node Bs involved in soft handover are controlled by another RNC, then this RNC is referred to as the Drift RNC (DRNC). The base stations in the active set try to execute closed-loop power control independently, which may result in the mobile terminal receiving conflicting power control commands. In this case, power-down commands have priority, since they imply that one base station in the active set receives the terminal's signal at sufficient quality.

With soft handover in the downlink direction, multiple base stations transmit to a single terminal in soft handover state. Whereas this is beneficial for the terminal, it leads to increased downlink interference. Therefore, a clear trade-off exists between the positive effect of macrodiversity and the negative effect of increased interference. The maximum net gain at equal pathloss is 2.5 dB. At a pathloss difference of 6 dB, a net loss of 0.5 dB is incurred; at a difference of 10 dB, the loss of event is 2.5 dB (Brand and Aghvami, 2002). Soft handover is only beneficial for a fraction of terminal population. In practice, the fraction of terminals in soft handover is below 30–40 % (Holma and Toskala, 2007).

3.2.2 Radio Access Network

The architecture of UTRAN is illustrated in Figure 3.1. The UTRAN consists of the RNC and the Node B or base station, which provides radio coverage to a cell. A Node B connects to the User Equipment (UE) via the Uu interface and to the RNC via the Iur interface. Logically, the UTRAN is subdivided into individual Radio Network Subsystems

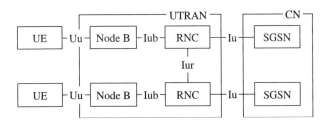

Figure 3.1 UTRAN architecture

(RNSs), where each RNS is controlled by an RNC. In Figure 3.1, the RNC is connected to the packet-switched domain of the CN.

The UE allows the user access to network services, with the radio link serving as the interface between the UE and the network. The UE itself comprises the User Subscriber Identity Module (USIM) and the Mobile Equipment (ME). The network maintains separate identities for the USIM and for the ME.

In UMTS, mobility management is performed hierarchically by the CN and by UTRAN. The CN tracks the location of the UE on the level of routing Areas (RAs). UTRAN-level mobility management refers to the functions performed at the RNC in order to keep the UE in touch with the UTRAN radio cells, taking into account the user's mobility within UTRAN and the type of traffic it is using (Kaarannen *et al.*, 2001). The RNC tracks the location of users with a Radio Resource Control (RRC) connection either on the level of a cell or on the level of a UTRAN Registration Area (URA). The URA is defined as an area covered by a number of cells. The URA is only known internally in UTRAN and hence is not visible to the CN. An RNC tracks a UE with an active RRC connection on cell level when data transfer is taking place and on URA level when no data is actively being transferred and the probability of data transfer is high (3GPP, 2002). If the UE is tracked on URA level, the RNC must perform the paging procedure in order to transfer data. We refer to both RRC states as cell connected and URA connected respectively.

Radio Interface Protocol Architecture

Three layers are relevant for the radio interface, namely the physical layer (layer 1 or PHY), the data link layer (layer 2) and the network layer (layer 3), the last two featuring several sublayers. Figure 3.2 shows the radio interface protocol architecture of UMTS. The lowest three layers and sublayers are uniformly referred to as the physical layer, the MAC layer and the Radio Link Control (RLC), with the MAC and RLC being sublayers of layer 2.

The radio interface protocol architecture is subdivided into the control plane (or C-plane) and the user plane (or U-plane). In the control plane, which deals with signalling, the RRC sits on top of the RLC. The RRC is the lowest sublayer of layer 3 and is the only sublayer of layer 3 fully associated and terminated in the UTRAN. In the user plane, additional sublayers may be required at layer 2, depending on the services supported. The Packet Data Convergence Protocol (PDCP) performs IP header compression and decompression, transfer of user data and maintenance of sequence

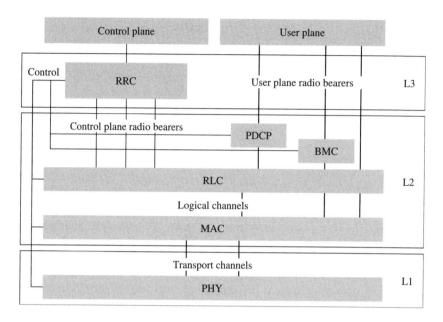

Figure 3.2 Protocol architecture of the WCDMA air interface (Brand and Aghvami, 2002)

numbers for radio bearers which are configured for lossless SRNC relocation. The Broadcast/Multicast Control (BMC) protocol is required for broadcast and multicast services.

The physical layer offers services to the MAC in the shape of transport channels, and the MAC to the RLC in the form of logical channels. A transport channel is characterized by how the information is transferred over the radio interface, while a logical channel is characterized by the type of information that is transferred. Layer 2 provides radio bearers to higher layers. The control plane radio bearers provided by the RLC to the RRC are signalling radio bearers. The RRC interfaces not only to the RLC but also to all other layers below it for control purposes (Brand and Aghvami, 2002). The reader is referred to (3GPP, 2008c) for further details on the radio interface protocol architecture.

Physical Layer Basics

The physical layer performs numerous functions as listed in (3GPP, 2007g) such as:

- macrodiversity distribution/combining and soft handover execution;
- FEC encoding/decoding of transport channels, error detection on transport channels and indication of errors to higher layers;
- multiplexing of transport channels onto so-called Coded Composite Transport Channels (CCTrCHs) on the transmit side, and demultiplexing from CCTrCHs to transport channels on the receiver side;
- mapping between CCTrCHs and physical channels;
- modulation/spreading and demodulation/despreading of physical channels;

- frequency and time synchronization, the latter on the level of chips, bits, slots and frames;
- measurement of radio characteristics including frame error rate, SIR, interference power and so forth, which are reported to higher layers;
- inner- or closed-loop power control.

In WCDMA, a 10 ms radio frame is divided into 15 regular slots. At a chip rate of 3.84 Mcps, each slot measures 2560 chips. The user data rate is kept constant during each radio frame. However, the data rate can change from frame to frame. Radio capacity allocation is typically controlled by the network to achieve optimum throughput for packet data services. The UTRA modulation scheme is Quadrature-Phase Shift-Keying (QPSK).

The physical layer makes use of physical channels for the delivery of data over the air interface. A physical channel is characterized by the code, the frequency and in the uplink also the relative phase, either I for in phase or Q for quadrature phase. The information rate for physical channels is variable, which is made possible by the use of variable Spreading Factors (SFs). As a result, the information rate of the channel is also variable. On the uplink, the spreading factors range from 256 to 4, and on the downlink from 512 to 4.

Signals are first spread using channelization codes, after which a scrambling code is applied at the same chip rate as the channelization code, such that the signal bandwidth is not altered by the scrambling code. Channelization codes are used to separate channels from the same source – on the downlink different channels in one sector or cell, on the uplink different dedicated channels used by a single mobile terminal. Scrambling codes are used to separate signals from different sources. The channelization codes are based on the Orthogonal Variable Spreading Factor (OVSF) technique, which allows mutually orthogonal codes to be chosen from a code tree, even when codes for different spreading factors are used simultaneously. In UTRA, two types of scrambling code are used, Gold codes with a period of 38 400 chips and so-called extended S(2) codes with a period of 256 chips, the latter optional and only applicable in the uplink.

The physical layer offers various types of transport channel to the MAC sublayer. Transport channels are unidirectional channels and can be classified into two groups: common transport channels that are shared by multiple users and feature inband identification of mobile terminals if a particular terminal is to be addressed, and dedicated transport channels, where, by virtue of a channel dedicated to a particular communication, the terminal is identified by the physical channel it uses (Brand and Aghvami, 2002).

The common transport channels supported in Release 99, the first release defining UMTS in FDD mode and the UTRAN, are as follows:

- the Random Access Channel (RACH) on the uplink;
- the Forward Access Channel (FACH) on the downlink;
- the Downlink Shared Channel (DSCH);
- the Common Packet Channel (CPCH) on the uplink;
- the Broadcast Channel (BCH) on the downlink;
- the Paging Channel (PCH), also on the downlink.

There is only one type of dedicated transport channel defined in Release 99, namely the Dedicated Channel (DCH).

The basic information unit delivered by the MAC on a transport channel to the physical layer is a transport block. Every so-called Transmission Time Interval (TTI), the MAC delivers either one or a set of transport blocks to the physical layer for a given transport channel. With a transport block set, all transport blocks are equally sized, with the block size changeable from TTI to TTI.

The characteristics of a given transport channel are determined by its transport format, with attributes such as the transport block size, the number of transport blocks in a transport block set, the TTI, the error protection scheme applied and the size of the Cyclic Redundancy Check (CRC). Transport channel characteristics can be defined in terms of a transport format set.

Layer 1 can multiplex one or several transport channels onto a coded composite transport channel, each of them with its own transport format picked from its transport format set. However, not all possible permutations of these combinations are allowed. Rather, only a set of authorized Transport Format Combinations (TFCs) may be used. On the transmit side, the physical layer builds the Transport Format Combination Identifier (TFCI) from the individual TFIs, which is then appended to the physical control signalling. By decoding the TFCI on the physical control channel, the receiving side has all the parameters needed to decode the information on the physical data channels and to deliver them to the MAC in the format of the appropriate transport channels.

MAC Layer Basics

As defined in (3GPP, 2008b), the functions performed by the MAC layer include:

- the mapping between logical channels and transport channels;
- the selection of appropriate transport formats for each transport channel, depending on the instantaneous source rate;
- various types of priority handling, both between data flows from one terminal or from different terminals;
- the identification of mobile terminals on common transport channels;
- the multiplexing of higher-layer Packet Data Units (PDUs) onto transport blocks to be delivered to the physical layer on the transmitting side, and demultiplexing of these PDUs from transport blocks delivered from the physical layer on the receiving side.

The MAC sublayer offers logical channels to the RLC. Logical channels can be classified into two groups, namely control channels for the transfer of control plane information and traffic channels for the transfer of user plane information. The control channels are:

- the Broadcast Control Channel (BCCH), a downlink channel used for broadcasting system control information;
- the Paging Control Channel (PCCH), a downlink channel used to transfer paging information, when the network does not know the UE location on cell level or when the UE is in sleep mode;
- the Common Control Channel (CCCH), a bidirectional channel used for transmitting control information;
- the Dedicated Control Channel (DCCH), a point-to-point bidirectional channel used for the transmission of dedicated control information between the UE and the network.

The two types of traffic channel are:

- Dedicated Traffic Channel (DTCH), a point-to-point uplink or downlink channel dedicated to one MS for the transfer of user information;
- the Common Traffic Channel (CTCH), a point-to-multipoint unidirectional (downlink only) channel used for the transfer of dedicated user information for all or a group of mobile terminals.

The DTCH can be mapped onto dedicated or common transport channels. This is owing to the distinction between the type of information transferred (as defined by the logical channel) and how the information is transferred over the radio interface at the level of transport channels.

Three different types of MAC entity are distinguished: the MAC-b, the MAC-s/ch and the MAC-d. The MAC-b handles the BCH at the network side, situated in the Node B. The MAC-s/ch handles all other common or shared transport channels. It is situated in the controlling RNC. Finally, the MAC-d handles the DCH, the only dedicated transport channel. The MAC-d is situated at the serving RNC.

RLC Layer Basics

The RLC layer provides three types of data transfer service to higher layers, namely transparent, unacknowledged and acknowledged data transfer. In the case of transparent data transfer, higher-layer PDUs are transmitted without adding any protocol information (for example, RLC headers). Unacknowledged data transfer means that higher-layer PDUs are transmitted without guaranteeing delivery to the peer entity. In unacknowledged mode, the RLC performs error detection and delivers only Service Data Units (SDUs) free of transmission errors to higher layers. In acknowledged mode, the RLC layer provides error-free transmission by means of appropriate Automatic Repeat Request (ARQ) techniques. ARQ is an error control method for data transmission that uses acknowledgments and time-outs to achieve reliable data transmission. An acknowledgment is a message sent by the receiver to the transmitter to indicate that it has correctly received a data frame or packet. If the sender does not receive an acknowledgement after the time-out period, it usually retransmits the frame or packet until it receives an acknowledgement or exceeds a predefined number of retransmissions.

Physical Channels

A physical channel is characterized by the code, the frequency and in the uplink also the relative phase, either for in-phase or Q for quadrature-phase. More precisely, the uplink modulation is a dual-channel QPSK, which means separate Binary-Phase Shift-Keying (BPSK) modulation of the I-channel and Q-channel. Downlink modulation is 'proper' QPSK (that is, a single channel is modulated onto both in phase and quadrature phase). It means that the symbol rates of an uplink and a downlink channel at a given spreading factor are the same, but that the downlink physical channel bit rate is double that of the uplink physical channel, for example 30 kbps as compared with 15 kbps at a spreading factor of 256. As well as physical channels, there are also physical signals, which do not have transport channels mapped to them. As usual, physical channels can be categorized as either dedicated or common physical channels.

All dedicated physical channels feature a radio frame length of 10 ms. In the uplink direction, a Dedicated Physical Control Channel (DPCCH) carrying layer 1 control information is code multiplexed with the Dedicated Physical Data Channel (DPDCH). In the downlink direction, there is effectively only one type of downlink Dedicated Physical Channel (DPCH), onto which data generated at layer 2 and above (i.e. the dedicated transport channel) is time multiplexed with layer 1 control information.

The common physical channels defined on the uplink are the Physical Random Access Channel (PRACH) and the Physical Common Packet Channel (PCPCH). Not surprisingly, they are used to carry the RACH and the CPCH respectively.

On the downlink, the following common physical channels and signals are defined:

- The Common Pilot Channel (CPICH), with exactly one mandatory Primary CPICH (P-CPICH) per cell and zero, one or several Secondary CPICHs (S-CPICHs). The P-CPICH must be broadcast over the entire cell, whereas the S-CPICH may also be transmitted over only a part of the cell, for example using smart antennas.
- The Primary Common Control Physical Channel (P-CCPCH), which is a fixed-rate channel used to carry the BCH.
- The Secondary Common Control Physical Channel (S-CCPCH), a variable-rate channel used to carry the FACH and the PCH.
- The Sychronization Channel (SCH), a downlink signal used for cell search, which consists of two subchannels, namely the primary and secondary SCH.
- The Physical Downlink Shared Channel (PDSCH) used to carry the DSCH.

Also part of the downlink common physical channels are a number of indicator channels. Four of them provide fast downlink signalling required for the operation of the uplink common physical channels, namely the PRACH and the PCPCH. The indicator channels are:

- the Acquisition Indicator Channel (AICH) used to carry Acquisition Indicators (AIs) responding to PRACH preambles;
- the CPCH Access Preamble Acquisition Indicator Channel (AP-AICH) carrying Access Preamble Acquisition Indicators (APIs) responding to CPCH access preambles;
- the CPCH Collision Detection/Channel Assignment Indicator Channel (CD/CA-ICH) carrying either Collision Detection Indicators (CDIs), or, if channel assignment is used for the CPCH, Collision Detection Indicators/Collision Assignment Indicators (CDIs/CAIs) in response to CPCH collision detection preambles
- the CPCH Status Indicator Channel (CSICH) signalling the availability of CPCHs through Status Indicators (SIs);
- the Paging Indicator Channel (PICH) carrying Paging Indicators (PIs).

Channel Mappings

The possible mappings between logical channels and transport channels are shown in Figure 3.3. The DTCH can be mapped onto common or dedicated transport channels,

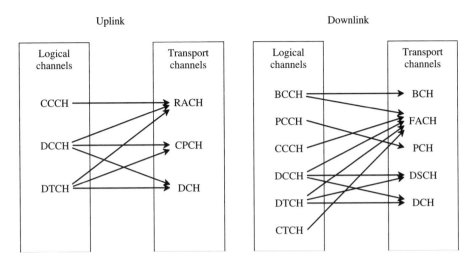

Figure 3.3 Mapping between logical and transport channels

hence onto the RACH, the CPCH, the DSCH, the FACH and the DCH. More than one DTCH can be mapped onto a single DCH, but different DTCHs can also be mapped onto different DCHs, depending on how the relevant radio bearers are configured.

The physical layer offers transport channels as services to higher layers. It also offers indicators, which are fast low-level signalling entities that can be transmitted without relying on information blocks sent over the transport channels. The mapping of transport channels and indicators to physical channels is illustrated in Figure 3.4.

3.2.3 Core Network

UMTS employs a backbone network for its radio access network based on IP. The UMTS Packet-Switched (PS) domain provides connectivity to external networks based on IP and the Point-to-Point Protocol (PPP) (Lin, Rao and Chlamtac, 2001).

The PS portion of the CN is illustrated in Figure 3.5 and consists of the GGSN and the SGSN. The GGSN interfaces to external Packet Data Networks (PDNs) via the Gi interface. An SGSN connects to the GGSN via the Gn interface and to UTRAN via the Iu interface. The GGSN can be seen as an edge IP router providing connectivity to IP networks (Koodli and Puuskari, 2001).

The Home Subscriber Server (HSS) is the master logical database and maintains user subscription information to control network services. The HSS encapsulates the Home Location Register (HLR), which maintains users' identities, locations and service subscription information. The HLR implements interfaces with the GGSN and SGSN.

Mobility Management

Mobility management enables the network to track the locations of users and their terminals between call or packet session arrivals (Wong and Leung, 2000).

Before a UE can exchange data with an external PDN, the UE must first establish a virtual connection with that PDN (Salkintzis, 2001). Once the UE is known to the network,

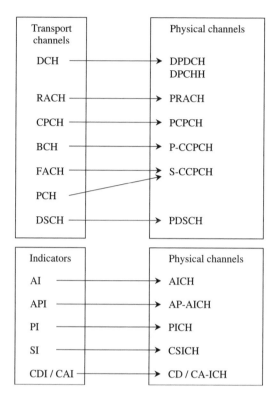

Figure 3.4 Mapping between transport channels and indicators to physical channels

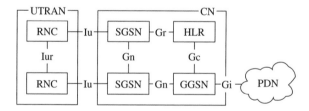

Figure 3.5 UMTS CN architecture

packets are transferred between the UE and the external PDN on the basis of the Packet Data Protocol (PDP), the network-layer protocol carried by UMTS. An instance of a PDP type is called a PDP context and contains all parameters describing the characteristics of the connection to an external network by means of endpoint addresses and Quality of Service (QoS) (Kaaranen *et al.*, 2001). A PDP context is established for all application traffic sourced from and destined for one IP address.

In UMTS, Mobility Management (MM) is performed hierarchically both by the CN and by the UTRAN. In the CN, an MM finite state machine is exercised in both the SGSN and the UE to characterize the mobility management activities for the UE (Lin *et al.*, 2001). The SGSN maintains MM contexts that store the mobility states of each user within its service area. The service area of an SGSN is partitioned into several RAs. When a UE moves from one RA to another, an RA update is performed, which informs the SGSN of the UE's current location (Lin and Chen, 2003). An MM context encapsulates the PDP contexts that are activated for a single subscriber. Similarly, the RNC maintains RNC contexts, each of which encapsulates zero or more RAB contexts. For the UMTS packet service domain, the Packet Mobility Management (PMM) states are PMM-DETACHED, PMM-IDLE and PMM-CONNECTED. In the PMM-DETACHED state, the UE is not reachable by the network. In the PMM-IDLE state, the UE is tracked by the SGSN on RA level and the paging procedure must be performed in order to reach the UE, for example for signalling (Kaarannen et al., 2001). Only in the PMM-CONNECTED state can packet data be transferred over the Iu interface, since only then does the SGSN have valid routing information for packet transfers with an accuracy of the routing address of the actual serving RNC. In the PMM-CONNECTED state, the location of the UE is tracked by the serving RNC (3GPP, 2007b).

In UMTS, a hierarchical tunnelling mechanism is used to transfer packets towards the mobile subscriber's actual point of attachment within the network. The network establishes tunnels between neighbouring network nodes along the path from the gateway to the mobile subscriber. Packets addressed to the mobile subscriber travel in the network of tunnels, which can be viewed as a separate routing network overlay on top of IP. This requires that network nodes maintain a list of subscriber entries and search this list for each downlink packet. Each entry contains a pointer to the next node towards the mobile subscriber's actual point of attachment (Campbell et al., 2002).

The activation of a PDP context is a request-reply procedure between a UE and the GGSN. A successful context activation leads to the creation of two GPRS Tunnelling Protocol (GTP) sessions specific to the subscriber: between the GGSN and SGSN over the Gn interface and between the SGSN and RNC over the Iu interface. IP packets destined for an application using a particular PDP context are augmented with UE- and PDP-specific fields and tunnelled using GTP to the appropriate SGSN. The SGSN recovers the IP packets, queries the appropriate PDP context, based on the UE- and PDP-specific fields, and forwards the packets to the appropriate RNC (Koodli and Puuskari, 2001). The RNC maintains Radio Access Bearer (RAB) contexts. Equivalently to PDP contexts, an RAB context allows the RNC to resolve the subscriber identity associated with a GTP-tunnelled Network PDU (N-PDU). The RNC recovers the GTP-tunnelled packet and forwards the packet to the appropriate Node B. A Tunnel Endpoint Identifier (TEID) is used across the Gn and Iu interfaces to identify a tunnel endpoint at the receiving network node.

Session Management

A UE with an active PDP context is said to be in the active Session Management (SM) state, whereas a UE without an active PDP context is said to be in the inactive SM state (Kaarannen et al., 2001). Prior to performing a PDP context activation, a PMM-IDLE user must perform the service request procedure, which establishes a secure connection to the SGSN. Once the service request procedure has been performed successfully, the terminal is able to send uplink signalling for the PDP context activation to the SGSN. The PMM and SM state transitions are shown in Figures 3.6 and 3.7 respectively.

The attach procedure allows the UE to be *known* by the network. For example, after the UE is powered on, the attach procedure must be executed before the UE can obtain

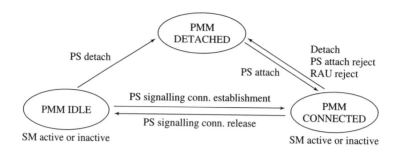

Figure 3.6 Mobility state transitions in UMTS (3GPP, 2007b)

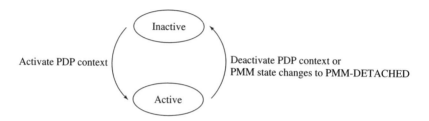

Figure 3.7 Session management state model in UMTS

access to UMTS services (Lin, Lee and Chlamtak, 2002). Similarly, the *detach* procedure is executed when the subscriber wishes to leave the network. The *detach* procedure may be performed implicitly when the network determines that the UE is not reachable, for example owing to battery removal or when the subscriber moves out of the service area. The network determines when to perform the implicit *detach* by maintaining a *mobile reachable timer* (3GPP, 2007b). Upon expiry of the mobile reachable timer, the SGSN moves the UE from the PMM-IDLE state to the PMM-DETACHED state.

Figure 3.7 shows the SM state parameters. The subscriber can be in the SM-active state when the corresponding PMM state is either PMM-IDLE or PMM-CONNECTED and the subscriber has activated a PDP context. A state transition from SM inactive to SM active takes place by means of the PDP context activation, whereas the opposite state transition occurs if the subscriber deactivates its PDP context or its PMM state changes to PMM-DETACHED.

3.3   Radio Access and Networking in CDMA2000

CDMA2000 is third-generation standard for mobile technology developed by the 3GPP2. The CDMA2000 standard has evolved continually to support new services in a standard 1.25 MHz carrier. The CDMA2000 standards are CDMA2000 1xRTT, CDMA2000 EV-DO and CDMA2000 EV-DV. CDMA2000 1xRTT provides data transmission capability with peak data rates at around 144 kbps. The data rate and cell capacity can be increased with either EV-DO or EV-DV. The data-only system EV-DO offers theoretical speeds of up to 2.4 Mbps. EV-DV offers both data and voice, but has yet to be deployed anywhere.

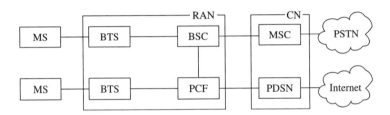

Figure 3.8 CDMA2000 architecture

As with UMTS, CDMA2000 networks consist of a core and radio access portion. The newly developed Packet Core Network (PCN) offers enhanced packet data services, whereas the radio access network portion of CDMA2000 is largely based on ANSI-41, the standard for intersystem signalling used in cdmaOne. The most important components of the core and radio access network of CDMA2000 1xRTT are illustrated in Figure 3.8.

The PCN for CDMA2000 relies on existing IETF networking standards such as mobile IP and consists of the Packet Data Serving Node (PDSN). The Packet Control Function (PCF) serves as the gateway to the radio access network. The PDSN establishes, maintains and terminates link-layer sessions, based on PPP to the MS (3GPP2, 2007b). The PDSN is therefore the mobility anchor point between the mobile node and the PCN. The PCF, on the other hand, is an abstraction for the radio access portion of the network and manages the relay of packets between the PDSN and the base station (3GPP2, 2007b). The PCF and PDSN are connected by an IP network. The Mobile Switching Centre (MSC) is responsible for most mobility management functionality such as handovers and paging in the radio access network and therefore still features strongly in the CDMA2000 architecture.

The radio access network consists of the Base Station Transceiver System (BTS) and the Base Station Controller (BSC). The BTS provides transmission capabilities across the air interface to the MS. The BSC provides control and management of one or more BTSs (3GPP2, 2002c). The BTS and the BSC combine into a single unit, the BS.

The architecture of the EV-DO network is similar to that of Figure 3.8, with the Access Terminal (AT) replacing the MS and the Access Network (AN) replacing the PCF.

3.3.1 Air Interface

CDMA2000 1xRTT is the core CDMA2000 wireless air interface standard, also known as IS-2000. The designation *1x* indicates the same bandwidth as cdmaOne, more specifically a duplex pair of 1.25 MHz radio channels. 1xRTT almost doubles the capacity of cdma2000, also known as IS-95, by adding 64 more traffic channels to the downlink or forward link, orthogonal to the original set of 64. Although capable of higher data rates, most deployments are limited to peak data rates of 144 kbs.

CDMA2000 EV-DO was designed as an evolution of the CDMA2000 standard that supports high data rates and can be deployed alongside existing cdmaOne voice services. An EV-DO channel has the same Radio Frequency (RF) bandwidth as CDMA2000 1xRTT. The channel structure, on the other hand, is very different. Additionally, the back-end network is entirely packet based, and thus is not constrained by the restrictions typically present in a circuit-switched network.

CDMA2000 EV-DV supports downlink data rates of up to 3.1 Mbs and uplink data rates of up to 1.8 Mbs. EV-DV can also support concurrent operation of legacy 1xRTT voice users, 1xRTT data users and high-speed EV-DV data users within the same radio channel.

As a result of equipment for EV-DV not being available in time to meet market demands, the EV-DV standard proved less attractive to operators compared with EV-DO, for which equipment and ASICs were available early on. In March 2005, Qualcomm suspended development of EV-DV chipsets and is now focused on improving the EV-DO product line.

3.3.2 Radio Access Network

The main radio access network functions in CDMA2000 include establishment, maintenance and termination of radio channels, as well as radio and mobility management. The radio access network for CDMA2000 EV-DO consists of the PCF and the BS.

The PCF is an abstraction for the radio access portion of the network and manages the relay of packets between the PDSN and the base station (3GPP2, 2007b). The PCF is a required IP element in CDMA2000 networks. It provides the relay of packets to the mobile terminal from the PDSN. It keeps track of the registration lifetime expiration and ensures that the session is renewed as necessary. It controls the available radio resources and buffers data received from the PDSN until radio resources become available.

The BS is further decomposed into the BSC and the BTS. The BSC controls one or multiple BTSs, which carry the radio-terminating equipment.

Radio Interface Protocol Architecture

Figure 3.9 depicts the architecture that applies to the family of CDMA2000 standards. The protocol architecture consists of the reference model layer 1, or physical layer, and layer 2, or link layer. In CDMA2000 terminology, *upper-layer services* refer to the OSI reference model layers 3 and above.

The physical layer provides coding and modulation services for a set of logical channels that are utilized by the multiplexing and QoS delivery sublayers shown in Figure 3.9. The physical layer is responsible for the modulation and coding of data traffic over the physical channel. Following coding and modulation, the physical layer generates a set of physical channels that are directly transmitted over the air. These physical channels can be broadly categorized into two basic classes, namely the Dedicated Physical Channel (DPHCH), the collection of all physical channels that carry information in a dedicated, point-to-point manner between the base station and a single mobile station, and the Common Physical Channel (CPHCH), the collection of all physical channels that carry information in a shared-access, point-to-multipoint manner between the base station and multiple mobile stations.

The link layer, or layer 2, is further subdivided into two sublayers, namely the Link Access Control (LAC) and MAC. The LAC sublayer provides transport of data over the air interface between peer upper-layer entities. It supports scalable and reliable transmission to meet the varying needs of the upper-layer entities. To provide this service, the LAC employs a number of different protocols to match the QoS requirements of each upper-layer entity to the characteristics of the MAC sublayer. For upper-layer entities that

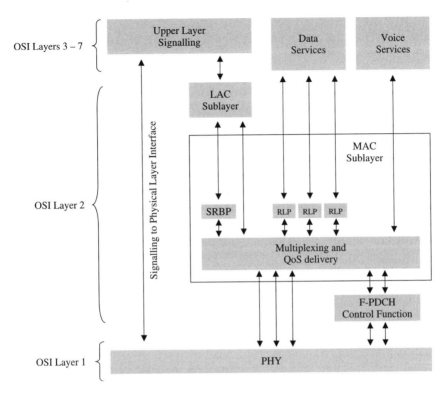

Figure 3.9 Protocol architecture of the CDMA2000 air interface (3GPP2, 2002b)

require a higher QoS than is provided directly by the MAC, the LAC enhances reliability through the use of various end-to-end reliable ARQ protocols that use sequence numbering, positive and negative acknowledgements and retransmission of lost or damaged packets. These protocols guarantee error-free delivery in sequence at the expense of added latency.

The MAC sublayer provides a control function that manages resources that are supplied by the physical layer and coordinates the usage of the resources desired by various LAC service entities. This coordination function resolves contention issues between LAC service entities within a single mobile station, as well as between competing mobile stations. The MAC sublayer is also responsible for delivering the QoS level requested by a LAC service entity. This can be done by either reserving air interface resources or by resolving priorities between competing LAC service entities.

The Signalling Radio Burst Protocol (SRBP) finally provides a mechanism for delivering signalling messages. The protocol is optimized for a common signalling channel and provides best-effort delivery.

The Radio Link Protocol (RLP) provides a best-effort streaming service. RLP provides both a transparent and a non-transparent mode of operation. In the non-transparent mode, RLP uses ARQ protocols to retransmit data segments that were not delivered properly by the physical layer. Non-transparent-mode RLP can introduce some transmission delay. In the transparent mode, RLP does not retransmit missing data segments;

however, RLP does maintain byte synchronization between the sender and receiver and notify the receiver of the missing portions of the data stream. Transparent RLP does not introduce any transmission delay, and is useful for implementing real-time services such as voice-over-RLP.

CDMA2000 EV-DO Revision 0

There have been several revisions of the EV-DO standard, starting with Revision 0. This was later expanded upon with Revision A to support QoS to enable operators to offer a variety of applications with different throughput and latency requirements. Revision B was published in 2006, with features that include the ability to bundle multiple carriers to achieve even higher rates and lower latencies. Revision B is a multicarrier evolution of the Revision A specification. It maintains the capabilities of EV-DO Revision A while providing higher rates per carrier. In this section, we briefly introduce EV-DO Revision 0. Further details on the EV-DO Revision 0 standard can be found in Bhushan *et al.* (2006).

As in cdmaOne and CDMA2000 1xRTT systems, the EV-DO Revision 0 carriers are allocated 1.25 MHz bandwidth and use a DS spread waveform at 1.2288 Mcps. The fundamental timing unit for downlink transmission is a 1.666 ms slot that contains the pilot and MAC channels and a data portion that may contain the traffic or control channel. Unlike CDMA2000 1xRTT, where a radio frame is 20 ms, a frame in EV-DO Revision 0 is 26.66 ms. Figure 3.10 illustrates the downlink slot structure, consisting of the pilot and MAC channels and the data portion that contains either the traffic or control channel.

The pilot channel is transmitted at full power for 96 chips every half-slot, providing a 1200 Hz sampling of the channel rate. These samples will be used to estimate and predict the received SIR at the AT, aiding the terminal in determining the maximum data rate that can be supported on the downlink. This provides the system with a mechanism for fast adaption of modulation and coding schemes to different mobile channel environments.

The MAC channel consists of a reverse activity channel and a Reverse Power Control (RPC) channel. The RA channel from a particular sector provides a 1-bit feedback to all terminals that can receive that sector's downlink, indicating whether or not its uplink load exceeds a threshold. The reverse power control channels from a particular sector carry a unique 1-bit closed-loop power control command (update rate of 600 Hz) for each of the ATs that include that particular sector in their active set. The Data Rate Control (DRC) lock channel is punctured into the RPC channel and is used to indicate the channel state from the AT to the access network. Figure 3.11 illustrates the downlink channel structure, including that of the MAC channel.

EV-DO Revision 0 uses a time-division multiplexed downlink. The traffic channel data rate used by the access network for transmission to an AT is determined by the DRC

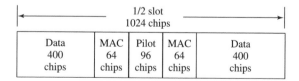

Figure 3.10   Downlink slot structure of EV-DO

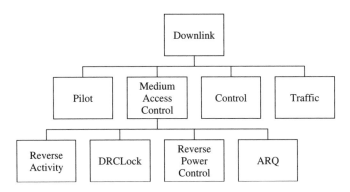

Figure 3.11   Downlink channel structure

message previously sent by the AT on the uplink. The DRC indicates not only the data rate but also modulation, code rate, preamble length and the maximum number of slots required to achieve the desired physical-layer error rate.

EV-DO Revision 0 introduces physical-layer Hybrid ARQ (H-ARQ) on the downlink. The access network transmits packets to an AT over multiple slots staggered in time. At higher SIR, significant coding gains are achieved by incremental transmission of parity bits, and, at lower SIR, powerful coding is achieved by simple repetition of a rate 1/5 turbo-encoded packet.

The EV-DO Revision 0 downlink traffic channel is a shared medium that provides high peak rate transmissions to active ATs. Addressing on the shared channel is achieved by a MAC index that is used to identify data transmissions from a sector to a particular AT. The packet preamble is covered with a biorthogonal sequence determined by the MAC index assigned to the AT. In order to maximize performance in a variety of channel conditions, three basic mechanisms exist to control access to the downlink traffic channel:

Open-loop rate control   A DRC message is sent by all ATs containing a requested data rate and a transmitting sector indication. The transmitting section chosen by the AT is the one that provides the best downlink channel and can receive the *a priori* downlink Channel State Information (CSI) with acceptable reliabilty.

Adaptive data scheduler   This takes into account fairness, queue sizes and the most recent *a priori* downlink CSI provided by the DRCs. While the specification does not specify the details of the scheduler, some form of proportional-fairness scheduler will typically be used in order to exploit multi-user diversity on the downlink.

Closed-loop rate control   A fast-feedback acknowledgement channel allows the data rate of a packet to be effectively increased beyond the data rate corresponding to the requested DRC if the channel conditions experienced by the transmission to the AT improve relative to the channel estimate used to generate the DRC.

The control channel in EV-DO Revision 0 is used for transmission of control and signalling information on the downlink. Two types of control channel, namely Sychronous Control (SC) and Asychronous Control (AC), are supported. The former is transmitted once every 256 slots, and the latter whenever needed, but never overlapping with the SC.

The AC may be used to transmit delay-sensitive signalling information to the ATs that either have an active connection with the access network already or are trying to establish an active connection. As an example, once the access network decodes an access probe from an AT, AC may be used to send an acknowledgement to the AT. On the other hand, using SC to send pages incurs large delays, as SC transmissions are rather infrequent. Moreoever, SC and AC are transmitted at 38.4 kpbs or 76.8 kbps using a 1024-bit payload to ensure high coverage for the control channel. However, this results in poor packing efficiency and inefficient utilization of downlink resources when transmitting a page to an AT using the SC.

3.3.3 Core Network

The PCN in CDMA2000 networks represents a first step in the evolution towards an all-IP and multimedia architecture, allowing for the delivery of enhanced packet data services. The PCN is a collection of logical and physical entities that provide IP-centric packet-data-based registration, roaming and forwarding services for mobile nodes (Murphy, 2001). The PCN consists of the PDSN, which serves as the mobility anchor point between the mobile node or MS and the PCN. The PCF and PDSN are connected by an IP network.

The initiation of a packet data service by the MS involves both the establishment of the required radio resources between the BS and the MS and the establishment of a PPP connection between the PDSN and the MS (3GPP2, 2002a).

Mobility Management

The PDSN is the point where the PPP connection to the MS is terminated. The states for a packet data service instance are inactive, dormant and active. Link-layer mobility management functionality is used for handovers between PCFs. This may happen when an MS is active or dormant and does not result in any change to the PPP session or IP addresses (3GPP2, 2001). Apart from establishing a new link-layer session between the new PCF and the radio access network and tearing down the session between the old PCF and the radio access network, a handover between PCFs may also require a handover between different PDSNs.

Mobility management procedures such as handover and paging for packet data services within the radio access network remain largely unchanged from the equivalent mobility management procedures in second-generation networks based on IS-95. Interestingly, handover signalling for inter-BS handovers as well as paging still involve the MSC (3GPP2, 2002a). This is in contrast to UMTS, where the MSC is not used in the PS domain.

In CDMA2000, mobility function support between different PDSNs within the home network is optional. The case where mobility between PDSNs is not supported is referred to as simple IP. With simple IP, the MS must obtain a new IP address when it moves to a new PDSN. In order to overcome such a break in IP-based connections and thus optionally to provide mobility functionality between PDSNs, CDMA2000 relies on Mobile IP (Perkins, 1996). With Mobile IP, mobile nodes use two IP addresses: a fixed home address and the CoA that changes at each new point of network attachment. The home address makes it appear that the mobile node is continually able to receive data on its home network. For this, Mobile IP requires the existence of an HA network node. Whenever the mobile node is not attached to its home network (and is therefore attached to what is termed a foreign network), the HA gets all the packets destined for the mobile node and arranges to deliver them to the mobile node's current point of attachment. Mobile nodes that are attached to a foreign network communicate with the FA. The FA operates by advertising its presence and services to any passing foreign users that have attached to its networks. Whenever the mobile node moves to a foreign network, it registers its new CoA, obtained from the FA, with its HA. To get a packet to a mobile node from its home network, the HA delivers the packet from the home network to the CoA, via what is known as an IP tunnel, in which all data is encapsulated within IP packets. The HA therefore acts as a proxy for the mobile user by taking the place of the home IP location and routing data to the FA, allowing communication with another host (Poole, 2004/2005). Handovers between two PDSN nodes in CDMA2000 rely on the standard Mobile IP procedures, as described in (Perkins, 1996).

Provided the optional IP-layer mobility function support between PDSNs is given, each PDSN contains the Mobile IP FA (McCann and Hiller, 2000). Handovers between PDSNs for MSs that are in the active state involve the establishment of a new PPP session, the detection of a new FA and the registration with the HA (3GPP2, 2001).

3.4 Summary

In this chapter, we have introduced the most important aspects of UMTS and CDMA2000 networks. For UMTS, the description of the WCDMA air interface covered the spreading and despreading operation, multipath reception by means of a Rake receiver, power control and soft handover. This was followed by an introduction to the radio interface protocol architecture, touching upon the physical, MAC and RLC layers. Finally, the channel mappings between the physical, transport and logical channels were depicted.

Similarly, for CDMA2000, the architecture of the radio access network as well as the radio interface protocol architecture were described. Several important aspects of the EV-DO Revision 0 standard were covered, with a focus on its channel structure. Finally, the architecture and several important procedures of the CDMA2000 core network were introduced.

4

Multicast Services for Third-Generation Networks

The purpose of this chapter is to provide an overview of mobile services that can be realized with multicast in third-generation networks. We first investigate the economic drivers behind introducing multicast. Then we describe several services that can be offered with multicast and formulate use cases for these services. With the use cases as a basis, we extract high-level requirements that the system must provide in order fully to support the described services. Finally, we analyse which factors have an influence on whether multicast services will be accepted by users and succeed in the marketplace.

4.1 Introduction

Mobile communications are evolving systematically from simple voice-centric point-to-point models towards more complex platforms offering a wide range of services. In the past, the structure of the communications industry was based on the tight integration of critical communication functionality. Services were provided by a limited set of players. The core of the network was designed to supply a tightly integrated set of features, with the users consuming these services at the edge of the network. With today's packet-based networks, service and delivery functionality is increasingly being decoupled.

Whereas today the most popular services are still communications oriented, such as voice and Short Message Service (SMS), mobile data services are increasingly gaining in popularity. Mobile data services are those services offered by network operators that provide subscribers with data while *on the go*. Mobile data services have significant potential, reflected by the number of diverse players all vying for the market. The value chain for mobile data services illustrates the diversity of players involved in providing such services and also highlights the decoupling of service and delivery functionality. A value chain is a chain of activities. A product or service passes through all activities of the chain in a particular order, and at each activity value is added to the product or

Multicast in Third-Generation Mobile Networks   Robert Rümmler, Alexander Gluhak and A. Hamid Aghvami
© 2009 John Wiley & Sons, Ltd

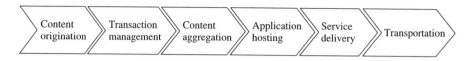

Figure 4.1   Value chain for the delivery of mobile data (MorganDoyle, 2001)

service. The mobile data value chain is an abstract classification of the different roles or functions that contribute to the overall value of a service delivered to mobile users. Figure 4.1 depicts the value chain for mobile data as it moves from the content originator to the consumer. Many real organizations or players can and will occupy several roles in the chain. The activities within the value chain are:

1. **Content origination** is the development of content such as music or information for electronic delivery over mobile networks. This function is generally performed by specialist producers of news, music and information services.
2. **Transaction management** includes the provision of clearing house functions for mobile electronic payments. A clearing house is a financial services company that provides clearing and settlement services for financial transactions.
3. **Content aggregation** includes mobile portal design, content hosting, processing of geographical and device information, personalization and content management functions.
4. **Application hosting** covers the hosting and management of applications such as universal messaging, e-commerce platforms, business applications and content delivery platforms.
5. **Service delivery** covers the customer acquisition and customer care functions associated with the value chain. This function is usually provided by either the mobile network operator or the mobile virtual network operator. A mobile virtual network operator is a company that provides mobile phone services but does not have its own licensed frequency allocation of radio spectrum, nor does it necessarily have all of the infrastructure required to provide mobile services.
6. **Transportation** is the provision of network connectivity for mobile communications. This role, by definition, is performed by the mobile network operator. This element adds value by shipping bits and bytes between end-users and other players. The user consumes the service through a service delivery player.

Two additional functions are required in order to deliver mobile data to the end-user, namely application and infrastructure development. Application development includes the development of applications such as games or business productivity tools for delivery over mobile networks. Infrastructure development covers the development and manufacturing of network components as well as end-devices used by customers.

Multiple roles can be taken up by the same players. The key players interact in the following way. The content owner owns the rights to sell and distribute content and cooperates with the content provider. The content provider is the party that provides the content for the services, with the content taking the form of either video, audio, text or binary data. The application provider builds the application around the content. The network operator owns the infrastructure for data transmission to the user, whereas the mobile service provider leverages the infrastructure of the network operator but offers its own services. The infrastructure manufacturer provides the technical infrastructure consisting of the network technology and the mobile terminals.

4.2 Motivation for Multicast

In the following, we highlight the mainly economic drivers behind the development of multicast technology in third-generation networks. Historically, network operators and infrastructure manufacturers have been the most instrumental players in developing third-generation technologies, with the other players taking a much less active role in driving new developments. We therefore focus on what network operators and infrastructure manufacturers stand to gain from multicast in third-generation networks.

4.2.1 Revenue Growth

Revenue is income that a company receives from its normal business activities. The revenue of a network operator is defined as the total amount of money it receives from sales of its services in a given period. Two quantities are used to calculate the total revenue of a network operator: the size of its subscriber base and the Average Revenue Per User (ARPU). The subscriber base is the primary (non-financial) measure for the size of its business. ARPU is defined as the average amount of revenue a network operator collects from each user per month and measures the value of the subscriber base. The total revenue is calculated by multiplying the size of the subscriber base with its ARPU.

In recent years, many network operators have recorded negative revenue growth for the first time. This is due to two reasons. Firstly, as a result of subscriber penetration reaching saturation, subscriber growth has been slowing. With penetration rates exceeding 100 % in many developed countries, continued subscriber growth is mainly due to subscribers owning multiple SIMs. Secondly, voice ARPU has been declining and will continue to do so because of price-based competition and lower mobile termination rates. As a result of this, total ARPU has been on the decline in most Western European countries.

Two avenues are being followed in order to reverse this trend: accelerating the pace of fixed-to-mobile substitution and enriching data services. Fixed-to-mobile substitution refers to the use of a mobile phone instead of a fixed telephone.

Markets with an underdeveloped fixed-line infrastructure offer extensive fixed-to-mobile substitution business expansion opportunities to operators, provided that low enough cost levels can be met to support local demand. In markets with well-developed fixed-line infrastructure, fixed-to-mobile substitution business opportunities leverage the fact that mobile communication has become an integrated part of a modern lifestyle.

With voice ARPU declining, network operators need to find ways to increase data ARPU. Enriching data services is key in reversing the trend of declining total ARPU. Currently, mobile messaging such as SMS still accounts for the lion's share of revenues from data services. Network operators need to adopt bold measures that will allow revenues from advanced data services such as those that can be realized with multicast to grow more quickly. Network operators are still very much focused on customers that are predominantly interested in voice. This has resulted in poor marketing that has deprived voice customers from accessing data services. Instead, network operators must now focus on the new brand of non-voice customers that are interested in accessing data services.

The revenue of a device manufacturer is defined as the total amount of money it receives from sales of its products in a given period. Mobile device suppliers are facing an uphill battle to increase revenues in the face of declining handset prices. Overall sales of mobile handsets worldwide have been levelling out. This is due to the saturation in the developed markets balancing out the booming growth that has been observed in developing regions.

A key motivation for device manufacturers is to increase revenue by developing and selling more user-friendly and interactive mobile devices that support the emerging mobile services such as those that can be realized with multicast.

4.2.2 Differentiation

In marketing, product differentiation or more simply differentiation is the process of distinguishing the differences of a product or offering from others, to make it more attractive to a particular target market. A successful product differentiation strategy will move a product from competing based primarily on price to competing on non-price factors, such as product characteristics, distribution strategy or promotional variables. This is done in order to demonstrate the unique aspects of a product and create a sense of value. The objective of differentiation is to develop a position that potential customers see as unique.

In the past, network operators could achieve differentiation by providing better coverage than competitors or by offering prepaid cards. Nowadays, offering novel services and unique content are the primary means of a network operator to achieve differentiation in the marketplace.

Device manufacturers, on the other hand, must differentiate themselves by producing user-friendly and reliable handsets that support the latest data services. Being *first movers* in bringing advanced handsets to the market highlights the technological leadership of a device manufacturer in comparison with its competitors. This is a key differentation strategy for device manufacturers.

4.2.3 Cost of Service Delivery

Network operators can increase their profit margins by reducing their operational expenses. One way to do this is by reducing their cost of service delivery. With the point-to-point delivery of multicast traffic, each user receives its own copy of data. This is not resource efficient. Resource efficiency can be greatly improved by means of multicast, as this allows multiple users to share the same network and radio resources. This also avoids the risk of capacity shortages when large groups of users are simultaneously interested in the same service offerings. As a result, network operators have scope to reduce their cost of service delivery by relying on multicast to deliver data to groups of users in a resource-efficient fashion.

4.3   Multicast Services

In this section, we give an overview of a number of services that may be realized with multicast. We define a service as a function provided by a service provider that is self-contained and does not depend on the context or state of other services. The multicast services we consider are grouped into six categories.

1. Mobile TV.
2. Multimedia Content Distribution.
3. General Content Distribution.
4. Enhanced Distribution Services.
5. Peer-to-Peer Communication.
6. Machine-to-Machine Communication.

We rely on use cases to document the multicast services. A use case defines a goal-oriented set of interactions between external actors (such as users) and the system under consideration. Use cases capture which actors do what with the system, for what purpose and towards what goal. The use case technique is used in software and systems engineering to capture the functional requirements of a system. We formulate use cases based on the intended usage of each of the multicast services. Each use case provides a brief description of the service, specifies the goal from the perspective of the user and lists the actors involved in delivering the service.

4.3.1 Mobile TV

Mobile TV refers to constant TV being provided to terminals. Mobile TV combines mobile technology with television content and represents a logical step both for operators, content providers and users alike. Mobile TV expands the universe of television by allowing television to be consumed on the go, no longer just in a stationary setting.

Mobile TV can also be delivered through one-way dedicated broadcast networks. There are several different standards or systems that allow mobile TV to be transmitted using broadcast technology. These include three primarily open standards developed by industry associations with contributions from multiple players in the mobile TV marketplace and one proprietary technology developed by Qualcomm.

The open standards developed by industry associations include Digital Video Broadcasting Handheld (DVB-H), Digital Multimedia Broadcasting (DMB) and Integrated Services Digital Broadcasting Terrestrial (ISDB-T). DVB-H stands out as an open standard specified by ETSI, but it requires new network infrastructure to be built and new spectrum to be allocated. DMB networks have been commercially launched in South Korea. The main disadvantage of DMB is that it is not supported by Nokia, the world's largest manufacturer of handsets. Nokia instead supports DVB-H. ISDB-T was adopted for commercial transmissions in December 2003 but is restricted to Japan only. MediaFLO is the proprietary technology developed by Qualcomm for broadcasting data to portable devices. The single biggest disadvantage of MediaFLO is its proprietary nature (with the resulting concentrated royalty fees) and the application of a vertically integrated business model. It is therefore unlikely to be implemented globally by operators.

With the infrastructure already in place and with sufficient capacity available, mobile TV is already being offered with third-generation mobile networks in unicast mode. Mass-market deployment of mobile TV has been prevented by bandwidth limitations, as unicast is not optimized to deliver the same content to many users at the same time. Delivering mobile TV in multicast mode on shared traffic channels will allow mobile TV to scale up for a mass market. Multicast is therefore necessary for large-scale access to mainstream TV channels through mobile devices. We consider two services for mobile TV:

1. Live TV.
2. Interactive TV.

Live TV

Live TV offers streaming access to traditional TV channels, with the added value of mobility. The user can choose from a selection of channels using an Electronic Programme Guide (EPG), an on-screen guide to scheduled television or radio programmes, with functions allowing a viewer to navigate, select and discover content. The average viewing time is likely to be shorter than with traditional TV. As with traditional TV, the earning logic is based on advertisements. In contrast to stationary TV, network operators can identify the individual users consuming live TV on their handsets. This provides marketing companies with the opportunity to develop new business models that are not possible with traditional TV. Three high-level user functionalities are required for live TV.

1. The user can select channels using a programme guide or portal.
2. The user can switch between channels.
3. The user can watch the video stream.

The use case for Live TV is described in Figure 4.2.

Interactive TV

Interactive TV gives users the opportunity to get involved with a TV show through voting. With this service, live TV is combined with a real-time feedback channel and potentially also user identification. Users pay per vote, with interactions being triggered as part of the live programme stream. The use case for interactive TV is described in Figure 4.3.

4.3.2 Multimedia Content Distribution

Multimedia content distribution covers the delivery of media-rich content to the user, in most cases on demand. On-demand services are services that offer content upon request. Typical examples are video and media-rich audio content. Multimedia content distribution may take place in a streaming or non-streaming fashion. With streaming, the content is consumed before the content has been received in its entirety, whereas, with non-streaming distribution, the entire download is performed prior to consumption. With multimedia content distribution, a back-channel may be required in order to allow for two-way interactions between the user and the service centre. Implementing such a back-channel

Name	Live TV
Category	Mobile TV
Description	The user selects a channel and watches TV on his mobile device
Goal	Allow users to watch TV while on the go
Actors	Content provider, network operator, user

Figure 4.2 Use case description for Live TV

Name	Interactive TV
Category	Mobile TV
Description	The user watches a TV show and has the opportunity to interact, for example by means of voting
Goal	Allow users to provide feedback during TV shows
Actors	Content provider, network operator, user

Figure 4.3   Use case description for Interactive TV

is trivial and can be done using the existing packet data mechanisms employed in third-generation networks.

We consider two multimedia content distribution services.

1. Video on Demand.
2. Media-Rich Radio.

Video on Demand

Video on demand allows subscribers to retrieve and watch a selection of movies at any time. Fixed video-on-demand systems either stream content through a set-top box, allowing viewing in real time, or download it to a device such as a digital video recorder for viewing at any time. As with mobile TV, extending video on-demand services to mobile devices is a natural progression and is technically feasible and economically viable with multicast technology. Video on demand requires a back-channel to allow subscribers to send movie selection information to the service centre. The use case is described in Figure 4.4.

Media-Rich Radio

Media-rich radio refers to adding non-audio content or interactivity to radio programmes. The term *visual radio* is often used to describe the addition of visual information to normal radio broadcasts. Nokia has trademarked a solution for visual radio, which relies on regular analog FM transmission for the radio programme and a synchronized and streamed data connection for the presentation of graphics and text.

We consider a media-rich radio service that is transmitted over the mobile network. The use case is described in Figure 4.5. With media-rich radio, information regarding artists or songs is distributed alongside the audio stream. Also, transactions such as voting, chatting

Name	Video on Demand
Category	Multimedia Content Distribution
Description	The users order video content and consume the content on their mobile devices
Goal	Allow users to access video content on demand
Actors	Content provider, network operator, user

Figure 4.4   Use case description for Video on Demand

Name	Media-Rich Radio
Category	Multimedia Content Distribution
Description	The user listens to a radio show which is enhanced with other media and also allows the user to interact
Goal	Allow users to obtain additional information or to interact with a radio show
Actors	Content provider, network operator, user

Figure 4.5   Use case description for Media-Rich Radio

or gaming can be offered to the user. Using multicast, media-rich radio is streamed over the mobile network, with a back-channel to the service centre enabling two-way interactivity.

4.3.3   General Content Distribution

General content distribution covers the distribution of general content to users, in most cases upon request. Typical examples are the distribution of information such as news, stock quotes, weather forecasts and the like. We consider three services in this category:

1. Data on Demand.
2. Content Casts.
3. Online Gaming.

Data on Demand

With the data-on-demand service, content is offered upon request. The user can request the download or streaming of content such as audio, video or text. A list of the content on offer is embedded in a portal. Users requesting the same data at similar times form a single multicast group. The delivery of the data is confirmed at the end of the service, with each user paying for the completed transaction. The use case for data on demand is described in Figure 4.6.

Content Casts

With content casting, users subscribe to periodically released content such as video podcasts, news or weather forecasts. This information is automatically sent to the user in a push fashion, using either Really Simple Syndication (RSS) or Atom syndication formats (Alexandri *et al.*, 2006). RSS is a family of web feed formats used to publish frequently

Name	Data on Demand
Category	General Content Distribution
Description	The user can order specific content such as audio, video or text
Goal	Allow users to download different types of data
Actors	Content provider, network operator, user

Figure 4.6   Use case description for Data on Demand

Name	Content Casts
Category	General Content Distribution
Description	The user subscribes to periodically released content
Goal	Allow users to stay up to date without having to request content proactively
Actors	Content provider, network operator, user

Figure 4.7 Use case description for Content Casts

Name	Online Gaming
Category	General Content Distribution
Description	Subscribed users are informed about game updates and receive these after service activation or automatically
Goal	Allow users to receive game updates in an automated fashion
Actors	Content provider, network operator, user

Figure 4.8 Use case description for Online Gaming

updated content in a standardized format. The use case description for content casts is given in Figure 4.7.

Online Gaming

Online games are games that require an online connection to exchange information with a central server. The data requested from the server can be game updates such as new levels or the high scores of other players. The use case description for online gaming is given in Figure 4.8.

4.3.4 Enhanced Distribution Services

Enhanced distribution services are services that require either a non-standard method of data transfer or reception or are specific to the given circumstances of a user. The services we consider in this category are the following:

1. Data Carousel.
2. Location-Specific Content Delivery.

Data Carousel

In a unidirectional transmission environment in which no back-channel is available, the receiver is unable to request the retransmission of data that was missed or received incorrectly. With the data carousel, the content of the transmission stream is provided in a cyclic fashion. The data carousel can be used to provide a 'ticker-tape'-type service in which data is provided to the user repetitively and updated at certain times to reflect changing circumstances. The use case description for carousel services is given in Figure 4.9.

Name	Data Carousel
Category	Enhanced Distribution Services
Description	Data is provided to the user repetitively and updated at certain times to reflect changing circumstances
Goal	Allow users always to have the most up-to-date data at any given time
Actors	Content provider, network operator, user

Figure 4.9 Use case description for Data Carousel

Name	Location-Specific Content Delivery
Category	Enhanced Distribution Services
Description	Location-specific information is transmitted to users within a given area
Goal	Allow users to receive location-specific information after subscription to a local channel
Actors	Content provider, network operator, user

Figure 4.10 Use case description for Location-Specific Content Delivery

Location-Specific Content Delivery

With this service, location-specific content is transmitted to users within a given location. The user can actively access location-specific information by subscribing to a local channel. The use case description for location-specific content delivery is given in Figure 4.10.

4.3.5 Peer-to-Peer Communication

A peer-to-peer network uses the diverse connections between participants within a network instead of relying on a relatively low number of servers to provide the core value to a service or application. This model of network arrangement differs from the client-server model where communication is usually to and from a central server. An important goal in peer-to-peer networks is that all clients provide resources, including bandwidth, storage space and computing power. Thus, as nodes arrive and the demand on the system increases, the total capacity of the system also increases. This is not true of a client-server architecture with a fixed set of servers, in which adding more clients could mean slower data transfer for all users. The distributed nature of peer-to-peer networks also increases robustness in the case of failures by replicating data over multiple peers.

Peer-to-peer networks are typically used for connecting nodes via largely ad hoc connections. Such networks are useful for many purposes. Sharing content files containing audio, video, data or anything in digital format is very common. Real-time data, such as telephony or messaging traffic, can also be passed between nodes using peer-to-peer technology. We consider two peer-to-peer communication services:

1. Data Carousel.
2. Location-Specific Content Delivery.

Name	Peer Content Distribution
Category	Peer-to-Peer Communication
Description	Users share data on a content platform within a network of peers
Goal	Allow users to share data within a peer network
Actors	Application host, network operator, user

Figure 4.11 Use case description for Peer Content Distribution

Name	Multiparty Conference Call
Category	Peer-to-Peer Communication
Description	Users communicate with a group of users in a conference call
Goal	Allow users to collaborate without the need for a physical meeting
Actors	Application host, network operator, user

Figure 4.12 Use case description for Multiparty Conference Call

Peer Content Distribution

With peer content distribution, users share audio, video or other data among the peer group. This service is very widespread in fixed networks, so that extending the service to mobile networks is a natural progression. The use case for peer content distribution is given in Figure 4.11.

Multiparty Conference Call

A conference call is a telephone call in which the calling party wishes to have more than one called party listen in to the audio portion of the call. Businesses use conference calls as a means to cut travel time and costs. Conference calls can take place over data networks, in which case multicast can be used to transfer data between the parties in the conference call. The use case for multiparty conference calls is given in Figure 4.12.

4.3.6 Machine-to-Machine Distribution

Machine-to-machine distribution refers to data communication between machines or computers connected through a network. In the context of mobile networks, machine-to-machine communication is particularly useful for upgrading or updating software, configuration data and new executable code on a device in order to modify its operation or performance.

Machine-to-machine distribution is triggered automatically or transparently by the device (acting on behalf of the user) or by a service provider. Machine-to-machine distribution is likely to involve a group of devices that, resulting from device commonalities such as the manufacturer, model name or software version, are targeted for an update or upgrade of existing functionality. These devices form a single multicast group.

Name	Software Distribution
Category	Machine-to-Machine Distribution
Description	The device of the user is updated automatically as soon as updates are available
Goal	Allow users to receive the latest software updates in an automated fashion
Actors	Infrastructure developer, network operator, user

Figure 4.13   Use case description for Software Distribution

Name	Navigation System Updates
Category	Machine-to-Machine Distribution
Description	The user receives regular navigation system updates
Goal	Allow users to receive navigation system updates in an automated fashion
Actors	Infrastructure developer, network operator, user

Figure 4.14   Use case description for Navigation System Updates

The services we consider in this category are:

1. Software Distribution.
2. Navigation System Updates.

Software Distribution

A drawback of existing wireless systems is their limited, or complete lack of, ability incrementally to upgrade functionality. Machine-to-machine distribution as a means of performing product enhancements, bug fixes and modest upgrade of end-user devices is of particular interest, since it is viable in terms of technology and has large economic potential (Blust, 2002). Device upgrades or updates that are transmitted to a group of targeted devices, for instance as a result of an important software upgrade or bug fix being made available by the device manufacturer, are of particular interest since such downloads greatly facilitate the maintenance of an installed base of subscriber units. The use case description for software distribution is given in Figure 4.13.

Navigation System Updates

Onboard navigation systems can also be updated dynamically with multicast. The updates apply both to system and map information as well as up-to-date traffic information or other localized content. This service is especially useful for built-in vehicle navigation systems. This requires an interface for navigation system vendors to the mobile network to allow them to update onboard navigation systems. The use case for navigation system updates is given in Figure 4.14.

4.4   User Requirements and Technology Acceptance

Although new technologies by themselves present possibilities, it is markets that decide whether or not to accept these new technologies (Picard, 1998). The acceptance of novel services in general is always determined by the extent to which they serve the wants and needs of users, the willingness of users to invest in the services, their willingness to pay service charges and their willingness to use their time differently (Picard, 2005). Thus, understanding user requirements and evaluating which factors most strongly affect user acceptance of new technology are critical in designing multicast services that can succeed in the marketplace.

In this section, we derive several high-level requirements that need to be fulfilled in order to deliver the services described in the previous section. We then investigate the adoption cycles of new technologies and look into the key factors that affect the user acceptance of new emerging technologies.

4.4.1 Requirement Analysis

Mobile services are highly complex products. Ideally, the design of a mobile service should take place in a highly structured manner, using a systems engineering approach. Systems engineering is an interdisciplinary field of engineering that focuses on the development and organization of complex systems. Using a waterfall model, Williams (2004) decomposes the development process in the following steps:

1. Analysing requirements.
2. Designing the system.
3. Implementing the system.
4. Testing the system.
5. Deploying the system.

The waterfall model for the development process according to Williams (2004) is illustrated in Figure 4.15. With the waterfall model, each of the steps in the development process are performed sequentially (Royce, 1970). The waterfall model maintains that one should move to a phase only when its preceding phase is completed and perfected. A number of modified waterfall models exist that include slight or major variations upon this process.

The analysis of requirements is typically the first step in the development process of a system. Requirement analysis encompasses those tasks that go into determining the needs or conditions that need to be met for a given product or system. Requirements can be

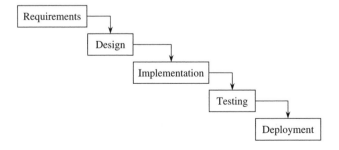

Figure 4.15   The waterfall development process model (Williams, 2004)

placed into two categories: functional requirements describe the functions that the system is to execute, whereas non-functional requirements are requirements that act to constrain the solution. Non-functional requirements are sometimes known as constraints or quality requirements.

In order to structure the most important requirements, we group the delivery of a multicast service into three distinct phases. The three service phases we consider are:

1. Pre-service phase.
2. Service delivery phase.
3. Post-service phase.

The functional and non-functional requirements defined for each of the three phases provide a high-level and non-exhaustive set of functionalities the system must provide in order to deliver the services described in the previous section.

Pre-Service Phase

The functional requirements during the pre-service phase are as follows:

1. **Service discovery**. The user must be able to discover which services are available within a given region or within the entire network.
2. **Subscription management**. Some services will require a subscription, whereas other services will not. The user must be able to subscribe for a given service as well as cancel a subscription at a later time.
3. **Service announcement**. The service announcement covers the announcement to a group of users that a service is about to commence. Service announcements can either take place selectively to a group of users that have subscribed to a service, to a group of users that are within a given service region or to all users within the network.
4. **Service activation**. The service activation covers the initiation (either as a push or a pull) and the start procedure for the service. The service activation can take place with or without a service announcement. Subscribed services may, for instance, take place periodically, without the need for an explicit service announcement.

Service Delivery Phase

The functional requirements during the service delivery phase, which may consist of one or more transmission phases, are as follows:

1. **Supported media types**. The following media types need to be supported: text, still images, video, speech and audio in mono and stereo. The media composition determines which protocols need to be employed for the transmission and the consumption of the services.
2. **Delivery method**. Some media types require streaming delivery, while other media types can be downloaded in their entirety prior to consumption. Streaming refers to the media delivery that is constantly received by, and normally displayed to, the end-user while it is still being delivered by the provider.

3. **Reliable delivery**. For many multicast services, reliability with respect to the delivery of data to the intended recipients is required. Without reliability, it cannot be guaranteed that data will be delivered intact, or that it will be delivered at all. Reliable delivery is especially important for file transfers, whereas for real-time traffic reliable delivery is typically not required.

4. **Carousel transmission**. For data carousel services, the user must be able to receive data that is provided repetitively and updated at certain times to reflect changing circumstances.

5. **Delivery verification**. For the distribution of critical data, a delivery verification may be required in order to ensure that all receivers have received the transmitted data correctly. This applies to the distribution of critical data such as files, software and other system updates.

6. **Interactivity**. For several services such as interactive TV and media-rich radio, the user is able to interact and provide feedback. This requires that a back-channel is implemented, which allows data to be transmitted from the device to the service centre.

The non-functional requirements for the service delivery phase are:

1. **Bandwith requirements**. Bandwidth requirements for multicast will range from 10 kbps for the distribution of lightweight data such as news up to 384 kbps for the distribution of video content. Audio streaming will require up to 48 kbps, while audio transmitted with low-quality video will require up to 128 kbps (3GPP, 2007f).

2. **Time-sensitive delivery**. Real-time traffic is very sensitive with respect to latency, defined as the delay of packet arrivals, and to jitter, defined as variations in latency. A very low amount of jitter is important for real-time applications using voice and real-time multimedia.

3. **Response times**. With services such as mobile TV, the response time in changing between channels must be comparable to the response times with stationary TV. Several tens of millseconds should be acceptable to most users.

Post-Service Phase

The functional requirements for the post-service phase are as follows:

1. **Service termination**. The service termination entails completing the delivery of service, either prematurely by the user or automatically at the end of the service.

2. **Charging**. Charging must be handled in such a way that any costs incurred are made transparent to the user.

4.4.2 Technology Adoption Cycles

New emerging technologies tend to have a high profile in the media and public debate compared with more mature technologies. This often leads to overinflated expectations, a so-called hype, around a particular technology. Linden and Fenn (2003) illustrate this phenomenon by hype cycles that aim at giving an overview of the relative maturity of technologies in a certain domain compared with their visibility.

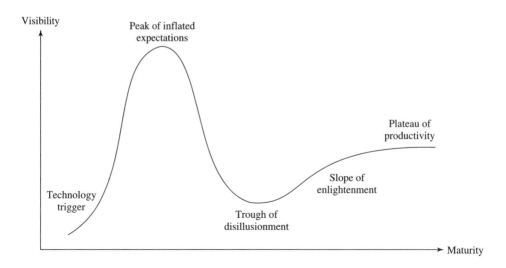

Figure 4.16   Hype cycle as defined by Linden and Fenn (2003)

A hype cycle is a graphic representation of the maturity, adoption and business application of specific technologies. The hype cycle characterizes the typical progression of an emerging technology from business and media overenthusiasm through a period of disillusionment to an eventual understanding of the technology's relevance and its role in a market or domain. Figure 4.16 illustrates the general structure of a hype cycle. Linden and Fenn (2003) define the phases of the hype cycle in the following way.

Technology trigger   The first phase of a hype cycle is the *technology trigger* or breakthrough, product launch or other event that generates significant press and interest.

Peak of inflated expectations   In the next phase, a frenzy of publicity typically generates overenthusiasm and unrealistic expectations.

Trough of disillusionment   Technologies enter the *trough of disillusionment* because they fail to meet expectations and quickly become unfashionable.

Slope of enlightenment   Some businesses or users continue through the *slope of enlightenment* and experiment to understand the benefits and practical application of the technology

Plateau of productivity   A technology reaches the *plateau of productivity* as the benefits of it become widely demonstrated and accepted.

Hype is generated by several actors: researchers, industry and media, pressing industrial actors for early launches of technologies. Launching immature technologies with incomplete business models then turns the hype curve into a steep descent. The new start with a more mature version of the technology is then characterized by a gentle upslope. Three adoption speeds for technologies can be distinguished (Linden and Fenn, 2003):

1. Fast-track technologies go through the hype cycle within 2 or 3 years. These technologies are typically adopted without many fanfares, bypassing the peak of inflated

expectations and the trough of disillusionment. High value to the users, simplicity of use, several strong vendors and use of current infrastructures are typical of fast-track technologies. SMS is perhaps the most famous example of a fast-track technology.

2. Long-fuse technologies may take one or two decades to traverse the hype cycle. Science-fiction-style fascination of the technology in the media, inherent complexity, reliance on a new infrastructure and required changes in business processes are typical of long-fuse technologies. Email and Internet, for instance, have gained their current success via a long-fuse hype cycle.

3. *Normal* technologies usually traverse the hype cycle in 5–8 years. These technologies usually introduce only small changes to existing user practices and technical infrastructures.

The typical timescale for a particular technology to be really mature and make a profit is 5–10 years after the first launch (Alahuhta, Jurvansuu and Pentikäinen, 2004). New technologies have often already been analysed, worn out and condemned in public debate before the technology is actually mature for large-scale deployment. As multicast is not an entirely new technology and only requires minimal changes to existing user practices and technical infrastructure, multicast is likely to have a fairly quick adoption speed, perhaps in the range of several years, thus falling into the fast-track bracket.

4.4.3 User Acceptance of Mobile Services

What are the factors that influence whether new technologies are accepted by users? Several models have been developed in the literature that attempt to isolate the most important factors that determine whether a technology will succeed in the marketplace.

The Technology Acceptance Model (TAM) developed by Davis (1989) explains the determinants of user acceptance of a wide range of end-user computing technologies, in particular for technology in the workplace. The model points out that perceived ease of use and perceived usefulness affect the intention to use. Perceived ease of use is defined as 'the degree to which a person believes that using a particular system would be free from effort', whereas perceived usefulness is defined as 'the degree to which a person believes that using a particular system would enhance his or her job performance' (Davis, 1989). Perceived ease of use also affects the perceived usefulness. The intention to use affects real usage behaviour. The model is illustrated in Figure 4.17.

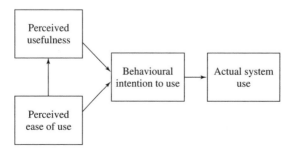

Figure 4.17   Original TAM (Davis, 1989)

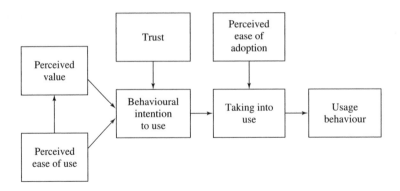

Figure 4.18   TAM for Mobile Services (Kaasinen, 2005)

Kaasinen (2005) studied the user acceptance of mobile services and developed the TAM for Mobile Services, a framework that identifies key factors that affect the diffusion of mobile technology to a wide user base. The TAM for Mobile Services extends the original TAM based on studies of mobile consumer services. Figure 4.18 illustrates the framework. The framework identifies the perceived value, perceived ease of use, trust and perceived ease of adoption as key factors in assessing the user acceptance of mobile services. The TAM for Mobile Services constitutes a solid framework for identifying issues that may affect user acceptance of technical solutions. In the following, we briefly describe how the perceived value, perceived ease of use, trust and perceived ease of adoption affect user acceptance and widespread adoption of mobile services.

Perceived Value

Consumers may lack a compelling motivation to adopt new mobile services unless those services create new choices where mobility really matters and manage to affect people's lives positively (Kaasinen, 2005).

Mobile services must create significant value for users. Value not only includes the utility of the service but also defines the key features that are appreciated by users. In a value-neutral setting, each requirement for a given service is treated as equally important in the design phase (Boehm, 2003). This easily leads to featurism, with a service becoming a collection of useful features but as a whole not providing enough value to the user. Focusing on perceived value in designing multicast services is therefore important and should occur in parallel to analysing the business or strategic value of a particular service.

Perceived Ease of Use

Davis (1989) defines perceived ease of use as 'the degree to which a person believes that using a particular system would be free from effort'. The perceived ease of use of a service is influenced in two ways. Indirect factors, such as the user's attitude towards technology in general, and experiences in using similar services, as well as information from other users, affect the initial perceived ease of use prior to having used the service. With actual and sustained use, perceived ease of use is directly affected by the user's own experiences of using the system in different contexts (Kaasinen, 2005).

In the case of mobile services, the limitations of the end-user device have a major influence on perceived ease of use. The limitations include the small screen, small and limited keyboard, the absence or limited functionality of pointing devices, limited amount of memory, limited battery power and slow connections. Designing mobile services for ease of use is to a large extent about coping with the limitations of the device.

Trust

Users are becoming more and more dependent on mobile services. Additionally, more and more personal data is involved in the relationship between the user and the service provider. As Gefen, Karahanna and Straub (2003) found in their studies of trust-related issues, the user's trust in the service provider is becoming a very important user acceptance factor.

Trust can be defined as an indicator of a positive belief about the perceived reliability, dependability and confidence in a person, object or process (Fogg and Tseng, 1999). Trust in mobile services includes perceived reliability of the technology and the service provider, reliance on the service in planned usage situations and the confidence that the user can keep the service under control and that the service will not misuse the user's personal data (Kaasinen, 2005).

Perceived Ease of Adoption

Users not being aware of the services that are available and not knowing how to take services into use constitute major obstacles in the widespread adoption of mobile services (Kaasinen, 2005). Perceived ease of adoption is defined as the degree to which a user believes that taking the service into use will be free from effort. As shown in Figure 4.18, perceived ease of adoption is positioned at the stage when the user's attention shifts from intention to use to actually taking the service into use.

Many mobile services require considerable configuration and personalization effort. This effort can overwhelm many users. Taking a service into use is a limiting factor in the transition from an intention to use to actual and sustained usage. Perceived ease of adoption is therefore a critical factor in achieving widespread adoption of mobile services. Service designers should therefore minimize the effort required in taking a service into use so that users can easily transition from an intention to use to actual and sustained usage.

4.5 Summary

A multitude of novel and promising services such as mobile TV and video on demand can be realized with multicast in third-generation networks. For network operators, the primary driver in introducing such services is to enrich their data offerings in order to increase data ARPU and thus reverse the trend of declining total ARPU as a result of price competition. In this chapter we have introduced several such services and derived the functional and non-functional requirements that must be met for the multicast services.

Technologies generally have different adoption speeds depending on whether new infrastructure is required and whether the technology requires changes to existing user practices. With multicast, the adoption is likely to take place fairly quickly, as only modest infrastructure investments are required and multicast services should be quite similar to the mobile data services already in use today.

It is clear that, if multicast services are to be successful, users must play a central part in the process of developing the services. Throughout the development process one must be able to answer questions such as: What will users get that they are not getting now? How is the technology or service relevant to their lives? How does it improve their lives or help them? Why is it valuable to them? Why should they use and pay for the new service? Whether multicast technology will be accepted by users and succeed in the marketplace greatly depends on a number of *soft* factors such as the perception of the technology in terms of ease of use and ease of adoption, whether it creates value for users and whether the user has sufficient trust in the service provider that the service will work reliably and will not infringe on the usage of personal data. All these factors need to be accounted for in designing multicast services.

5

Multicast Extensions for Third-Generation Networks

5.1 Introduction

This chapter introduces the architectural extensions that are required to support the efficient delivery of broadcast and multicast services in UMTS and CDMA2000 networks. The MBMS and BCMCS standards encompass the architectural extensions for multicast in UMTS and CDMA2000 networks respectively. The standardization process for MBMS and BCMCS, performed in the respective working groups of the 3GPP and 3GPP2, has created a phlethora of technical specifications that address different aspects of the multicast standards. The set of specifications provides the technical details required to implement network and end-user equipment such that network components are compliant with the standard and interoperable with one another.

We introduce the architecture and key concepts of both MBMS and BCMCS, alongside the most important mechanisms and protocols. References to the relevant technical specifications for MBMS and BCMCS are also provided, allowing the reader to obtain additional information and details on the standards. The chapter is structured as follows. Section 5.2 describes the main concepts of MBMS and gives details on the architectural extensions and the different service provisioning phases. Section 5.3 describes these aspects for BCMCS. The chapter closes with a discussion of the commonalities and key differences between both standards.

5.2 MBMS for UMTS

UMTS networks offer a highly sophisticated network bearer architecture for data delivery to mobile users in a cellular environment. In UMTS, the common packet-domain CN is used to interface to different RANs. On top of this IP-based infrastructure, UMTS implements an overlay architecture providing authentication and charging, mobility management and packet forwarding.

Multicast in Third-Generation Mobile Networks   Robert Rümmler, Alexander Gluhak and A. Hamid Aghvami
© 2009 John Wiley & Sons, Ltd

UMTS was initially designed for point-to-point communication between two parties. Until Release 6, the reception of IP multicast data was supported as an option. With this option, IP multicast terminates at the GGSN. Multicast applications running on a UMTS mobile terminal can use a group management protocol to join a multicast group via the GGSN. The GGSN acts as a multicast-enabled router and delivers incoming multicast traffic via point-to-point connections to interested receivers. With this solution, no bandwidth savings are achieved within the UMTS network. Data is delivered via individual connections to each receiver, instead of using shared resources in CN and the RAN.

MBMS was introduced in Release 6 in order to support resource-efficient delivery of multicast traffic in UMTS networks. MBMS provides point-to-multipoint bearer services, which can be used by multicast or broadcast applications to transfer data efficiently from one source to multiple mobile receivers. MBMS supports the delivery of both IPv4 and IPv6 multicast data and makes use of existing IETF mechanisms whenever possible and appropriate. In the specifications, MBMS bearer services are also often referred to as MBMS transport services. In the following, an overview of the functional MBMS architecture is provided.

5.2.1 Overview of MBMS Architecture

MBMS provides resource-efficient multicast bearer services that can be used by multicast user services or applications to deliver service content to a potentially large number of receivers. MBMS bearer services use shared delivery paths in the core network up to the radio access network and separate or shared downlink channels over the radio link, depending on the number of users in a given cell. Two different modes are defined for the MBMS bearer service, namely a broadcast and a multicast mode.

The broadcast mode is intended for sending service data from a single source to all receivers in a predefined broadcast area. A broadcast area can be preconfigured and may consist of one or several UMTS cells. In the multicast mode, service data can be delivered from a source to a multicast group in a multicast service area. Similarly to the broadcast mode, the multicast service area can also be preconfigured, but data is only delivered to those cells in the service area with multicast users that have expressed interest in receiving the service. Both modes intend to make efficient use of radio and network resources. Broadcast and multicast services can consist of either a single ongoing session or several intermittent sessions. Unlike in broadcast mode, where every user in the service area is able to receive the MBMS data, multicast mode requires subscription to the multicast group and users joining the group prior to the start of the session. This allows the network to charge users for a multicast session.

Figure 5.1 shows the UMTS reference architecture, including the functional entities required by MBMS. Besides a new network entity referred to as the Broadcast/Multicast Service Centre (BM-SC), MBMS also extends the existing nodes such as the GGSN, SGSN and the RNC with MBMS-specific functionality. Similarly, MBMS also requires additional functionality in the protocol stack of the UE. The following two sections describe these extensions in more detail.

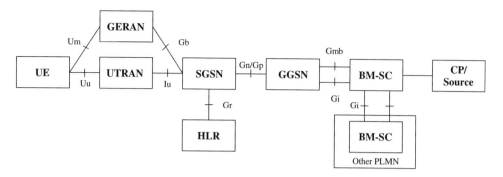

Figure 5.1   MBMS reference architecture

5.2.2   Core Network Extensions

The most apparent MBMS-related change to the UMTS architecture is the introduction of the BM-SC. The BM-SC is the entry point for a content provider to utilize the MBMS services inside the Public Land Mobile Network (PLMN), the term used to refer to a UMTS network in the 3GPP specifications. The BM-SC authenticates and authorizes content providers and verifies the integrity of the content. It can determine the QoS parameters of the MBMS service, allows the definition of the service area and generates charging data for the content provider. It also provides functions to announce the service and to schedule the MBMS data for transmission. The Gmb reference point has been added to provide an interface for control plane signalling between the BM-SC and the GGSN. IP multicast data is delivered to the MBMS bearer services via the already existing Gi reference point, which provides the interface between the GGSN and public data networks.

In order to support roaming of service users from other PLMNs, a roaming variant of the Gmb reference point, referred to as the Mz reference point, has been introduced. Signalling between the home and visiting BM-SC takes place via this interface whenever an MBMS service of the home PLMN is offered in the visiting PLMN.

Figure 5.2 shows a detailed overview of the service function of the BM-SC and its respective interfaces. As can be seen from the figure, the functionality of the BM-SC encompasses the following:

- membership function;
- session and transmission function;
- proxy and transport function;
- service announcement function;
- security function.

The *membership function* handles the subscription management for MBMS service users. It is an MBMS bearer service function that authenticates UEs that request the activation of a particular MBMS service. In the case where the membership function acts as a user service level function, it also implements the Gi interface. This requires

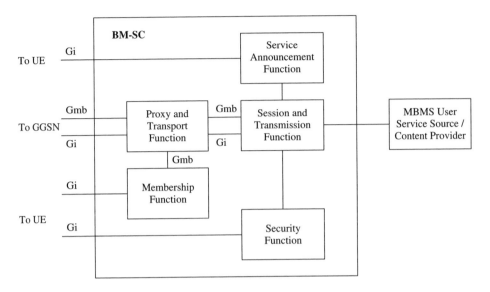

Figure 5.2   Service funtions of the BM-SC

the membership function to manage subscription data and generate charging records for MBMS service users.

The *session and transmission* function is a user-level service function that is responsible for the scheduling of MBMS sessions. It authenticates and authorizes content sources for MBMS sessions. It is also responsible for requesting the establishment and release of MBMS bearer resources in the network required for a particular MBMS session. For MBMS sessions that require reliable data delivery, the session and transmission function may apply error-resilient schemes such as FEC to the user service data. The MBMS also manages retransmissions, if necessary.

The *proxy and transport function* is an MBMS bearer service function that acts as a proxy agent for signalling over the Gmb reference point between GGSN and other BM-SC service functions. It allows for easy exchange of service data and necessary control signalling between the GGSN and a BM-SC that has its service functions distributed over multiple physical network elements. The proxy and transport function is also responsible for the generation of charging records for content providers.

The *service announcement function* is a user-level service function that is, as its name implies, responsible for the announcements of multicast and broadcast user services. The main task of this service function is to provide the UE with the necessary information required to join upcoming multicast sessions. This includes the media descriptions specifying the media to be delivered as part of an MBMS service, for example audio and video codecs, as well as MBMS session descriptions that identify session transmission time and addressing information for the MBMS user service. The announcement and description of sessions can be delivered via IETF protocols such as SDP, which is described in Chapter 2. For the delivery of session announcements, a variety of different network mechanisms can be utilized, including, but not limited to, MBMS bearer services or other push mechanisms such as WAP or SMS.

Finally, the *security function* is a user-level service function that provides a variety of different integrity and confidentiality protection mechanisms. These mechanisms can be used by multicast user services to protect the MBMS user data for transmission. The security function can also handle the distribution of encryption keys to authorized UEs.

In addition to the introduction of the BM-SC, the MBMS specification requires extensions to both the SGSN and GGSN to handle the management of MBMS bearer services. The GGSN serves as entry point for IP multicast traffic for MBMS data. The GGSN provides the service mechanism to initiate the establishment and release of an MBMS bearer plane towards the SGSN for a broadcast or multicast transmission. This process is usually triggered before or after an MBMS session by the BM-SC. For each MBMS bearer service, additional state is maintained within the GGSN. This state is referred to as an MBMS context and is required for the correct handling and forwarding of incoming MBMS service data within the UMTS network and the collection of charging records. New service procedures for the GGSN have also been added in order to manage MBMS sessions over the Gi interface.

The SGSN is responsible for the MBMS bearer service management of each individual UE, allowing the SGSN to transmit MBMS service data to the relevant radio access network segments within the UMTS network. As for the GGSN, additional state in the form of an MBMS context is required in the SGSN. This state is necessary for the correct handling and forwarding of incoming MBMS service data and the generation of charging records per UE for each MBMS bearer service. The SGSN usually keeps track of the capabilities of each UE, required for the establishment of adequate bearer paths, initiates the establishment and release of MBMS bearer paths towards the relevant radio access network segments and handles the exchange of MBMS contexts in the case of inter-SGSN mobility. Details of the different MBMS contexts and management procedures are described in Chapter 6. Alternatively, the reader can also consult the technical specification in (3GPP, 2007d) for further details.

5.2.3 Radio Access Network Extensions

The introduction of MBMS has also made changes to the UTRAN protocol architecture necessary. One of the main architectural design requirements was to minimize the impact on UTRAN by reusing as much as possible the UTRAN physical layer and other UTRAN mechanisms. In the following, a brief overview of the UTRAN extensions for MBMS is provided. Further details can be found in (3GPP2, 2008a).

In order to support a new point-to-multipoint bearer service for multicast delivery over the Uu interface, a new logical channel structure has been specified. Three new logical channels have been introduced for point-to-multipoint transmissions:

- MBMS Traffic Channel (MTCH);
- MBMS Control Channel (MCCH);
- MBMS Scheduling Channel (MSCH).

The MTCH is used for the downlink transmissions of user plane information between UTRAN and the UE. The user plane information is the service data that is sent over one MBMS bearer service. There is typically one MTCH per active MBMS bearer service. The MTCH is only sent in cells with an activated MBMS service.

The MCCH is used for the downlink transmission of control information between UTRAN and the UE. The MCCH carries control information specific to an MBMS bearer service. The MCCH can carry data such as information on neighbouring MBMS cells, information on MBMS services and the MBMS radio bearer. As for the MTCH, the MCCH is only sent in cells with an activated MBMS service.

The MSCH is used for the downlink transmission of an MBMS service transmission schedule between UTRAN and the UE. It allows UEs to perform discontinuous reception of an MTCH, thus saving receiver energy. The MSCH typically carries information such as MBMS service identities or the start time and duration of a period of data transmission. The MSCH is always transmitted in cells in which an MTCH is transmitted. There is always one MSCH per utilized physical-layer channel. Critical changes to information sent on the MCCH are indicated to the UE via MBMS notification indications. These MBMS notifications are transmitted via a newly introduced physical channel, the MBMS Notification Indicator Channel (MICH). The MICH represents an MBMS-specific channel very similar to the Paging Indicator Channel (PICH).

All three logical channels are mapped to the Forward Access Channel (FACH). The FACH in turn is mapped onto the Secondary Common Control Physical Channel (S-CCPCH). In order to support the new logical PTM control and user plane channels, the UTRAN MAC architecture (MAC-c/sh) has been extended with multicast functionality. The MBMS-specific MAC entity is referred to as the MAC-m. On the UTRAN side, the new MAC-c/sh/m entity is located in the controlling RNC, while at least one corresponding MAC-m is located in the UE. Figure 5.3 shows an overview of the resulting new UTRAN MAC architecture for MBMS transmissions. The MAC-c/sh/m entity performs scheduling, buffering and priortization of both MBMS and non-MBMS data flows. The addition of an MBMS-ID is required to distinguish between different MBMS services or, more specifically, the MTCHs that are mapped to the same transport channel.

The use of multiple dedicated or Point-to-Point (PTP) channels over the air interface is often more resource-efficient than the use of a single or shared Point-to-Multipoint (PTM) channel. Also, PTM channels such as the FACH do not support closed-loop power control, whereas PTP channels such as the DCH usually do. The MBMS standard therefore supports two transmission modes, namely PTP or PTM transmissions. One important factor in the selection of either of the two modes is the number of users in a cell. The optimum point for the selection is also different for each operational case, depending on additional factors such as the network configuration or the users' positions in a cell. Therefore, the MBMBS standard does not define a threshold number for the reselection between the two modes and leaves this to the operator as an implementation choice.

The RAN segment is not aware of the exact number of service users that are interested in receiving a particular MBMS multicast service. In order to overcome this limitation, the MBMS standard defines a function referred to as *counting*. This procedure allows a subset of UEs in a cell to report that they have joined a particular MBMS multicast service. Based on the results of the counting procedure, the network can decide whether the PTP or PTM transmission mode is most appropriate. The size of the subset can be customized by configuring the probability of a UE responding to the counting request. Having each receiver of a multicast service respond to the counting request is not desirable, as it may lead to an unnecessary high traffic load and congestion for larger receiver groups.

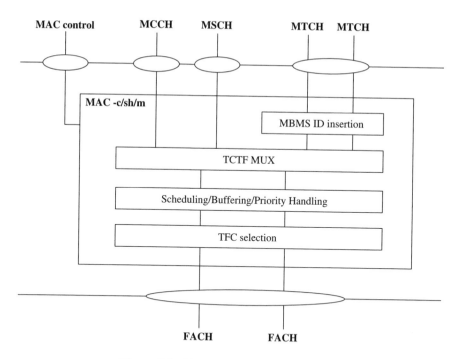

Figure 5.3 New MBMS MAC in UTRAN

A multicast receiver can receive the same service data from multiple radio links and combine this information, given that the same MBMS service data is transmitted in multiple adjacent cells. This type of macrodiversity can reduce the block error rate at the cell boundaries, thus allowing the required transmission power to be reduced and in doing so increasing the overall cell capacity.

The MBMS standard describes two methods for macrodiversity, namely selective combining and soft combining. With selective combining, a UE receives and simultaneously decodes packets from radio links of multiple adjacent cells/sectors. The received packets are then handled at the RLC layer. The RLC layer keeps the first correctly received packet and discards all other packets. A packet that is erroneously received on one link may be easily recovered by a correctly received packet from another link. In contrast, soft combining mitigates the reception of an erroneous packet on the physical layer. Incoming packets are directly combined on the physical layer and decoded after combination. Aho *et al.* (2007) provide more details on the performance of both schemes.

The MBMS standard also defines procedures for the management of the Iu bearer plane between the SGSN and the RNC for MBMS radio access bearers, as well as introducing an MBMS service context in the RNC. Details on these procedures are provided in Chapter 6.

5.2.4 Multicast Service Provisioning Phases

As a result of different service mechanisms, MBMS bearer services in multicast or broadcast mode have slightly different service provisioning phases. This section examines the provisioning of MBMS bearer services in multicast mode.

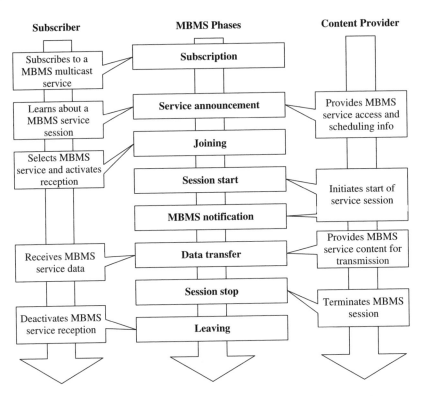

Figure 5.4 Service provisioning phases for MBMS multicast services

Figure 5.4 illustrates the eight different provisioning phases for multicast services. Some of the phases directly involve the service user, whereas others do not. A user wishing to receive a particular MBMS multicast service must first subscribe to the service. The subscription establishes a relationship between the user and the service provider in the form of a service agreement and is obtained during the subscription phase. The subscription information is recorded in the BM-SC, and only if a subscription exists will the operator provide the MBMS service to the user. The standard does not specify how subscriptions can be performed. Network operators and service providers can use any number of available methods such as establishing a subscription electronically over a web portal or by signing a physical contract.

The service announcement phase typically starts well before an MBMS service session and gives the service provider the opportunity to advertise an upcoming session to potential customers. The reception of service announcements is not limited to users with existing subscriptions. Service announcements can carry information such as a description of the service, its content and the media types that apply to the service, the start time and duration of the service and other information relevant for joining the service session. Service announcements can be distributed by MBMS, SMS or Multimedia Messaging Service (MMS) or by advertisements on websites.

Once a potential service user is informed about an upcoming service session, the user may opt to join the service session. Non-subscribers need to obtain a subscription prior to doing so. The joining process initiates the establishment of the MBMS multicast bearer plane. Joining usually takes place immediately before the session start, however MBMS bearer services that are configured to be *always on* may allow joining well before the session start, for instance at the time of receiving the service announcement. The joining phase is also referred to as the MBMS multicast service activation in the MBMS specification.

The session start phase is triggered by the BM-SC when session data is ready to be sent. The session start is transparent to the service users and initiates the reservation and establishment of bearer resources in the network for the imminent data transfer. Once all necessary bearer resources are established, the data transfer phase can commence. During the data transfer phase, MBMS multicast service data is transmitted to all receivers that have subscribed and joined or are in the process of joining the group during the multicast service session. MBMS notifications are usually sent during an ongoing MBMS data transfer in order to inform service users of the status of upcoming or ongoing MBMS multicast data transfers.

At any point in time the user may choose to leave a multicast session. Leaving can take place during a session, after a session or even before a session has started. The leaving process removes the service user from the multicast group and indicates to the UMTS network that the user no longer wishes to receive data from the indicated multicast bearer service. This is important as the network can save network bearer resources in areas in which no interested service users are present. This leave process is also referred to as multicast service deactivation.

Finally, the session stop phase is triggered by the BM-SC when the multicast session is terminated and no more session data is available for transmission. The session stop initiates the release of MBMS multicast bearer resources. The session stop can also be triggered by the BM-SC before the official end of a service session for sufficiently long periods of inactivity that justify the release of bearer resources. In such cases, the session can be resumed by a session start phase as soon as data becomes available at the BM-SC.

Figure 5.5 illustrates a typical timeline for the different multicast service provisioning phases. The horizontal axis represents time. The first three horizontal lines from the top of the figure represent events, data transmission activity for service announcement and MBMS service data respectively. These three horizontal lines apply to an example multicast service A. The remaining horizontal lines 4 to 7 illustrate events and data transmission activity for two service users UE1 and UE2 respectively. UE1 and UE2 are assumed to be located in different cells.

Following the event timeline of the multicast service A, the service provider first schedules a service announcement for an upcoming session of multicast service A. As illustrated in the figure, the service announcement lasts for the entire duration of two service sessions and is terminated only after the end of the second session. The data transfer of both service sessions does not begin directly after a session start, but instead after a short idle period. This is important, as data should only be sent once the required multicast bearer resources have been established. The first service session has an intermediate period of

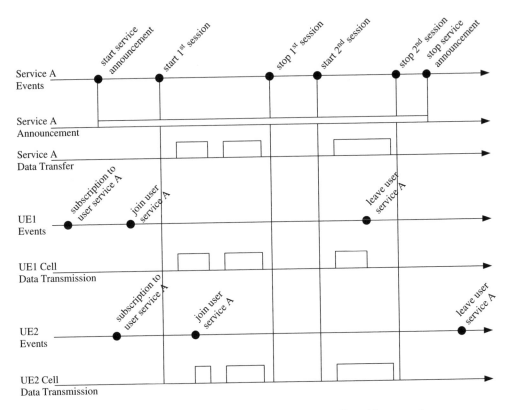

Figure 5.5   Example delivery timeline for a MBMS multicast service

data inactivity, in which the multicast bearer resources are released by a session stop phase and resumed by a session start phase. In the second service session, data transfer is continuous for the entire duration of the session.

UE1 subscribes to the multicast service well before the service announcement begins and decides to join the service session once it has been informed of the availability via the service announcement. As expected, the UMTS network transmits the service data for multicast service A in the cell in which UE1 is located. Halfway through the second session, UE2 decides to leave the multicast service by performing a multicast service deactivation. In this example it is assumed that UE1 is the last service user for the multicast service A in its network cell. As a result, the UMTS network stops transmitting service data in the cell.

UE2 has not previously subscribed to the multicast service A but decides to do so after being informed of the upcoming multicast session from the service announcement. UE2 joins the multicast service during the first service session. It is assumed that no previous service users for multicast service A are present in the cell in which UE2 is located. As a result, the data is transmitted in the network cell only after UE2 has activated the multicast user service A. UE2 leaves the multicast service only after the end of the second session and therefore receives the service data for the entire duration of the second session.

5.2.5 Broadcast Service Provisioning Phases

This section introduces the service provisioning phases of MBMS in broadcast mode. The provision of MBMS services in broadcast mode is less complex than the provision of MBMS services in multicast mode owing to the use of simpler service mechanisms for bearer management. In multicast mode, the UMTS network needs to be aware of every service user that is receiving an MBMS service. This is not required in broadcast mode. As a result, only content providers can be billed for the usage of MBMS broadcast bearer services, and every receiver is able to receive a broadcast service without requiring prior subscription.

Figure 5.6 shows the service provisioning phases for MBMS in broadcast mode. As can be seen from the figure, fewer phases exist compared with service provisioning of MBMS in multicast mode.

The subscription phase used for service provisioning in multicast mode is not required in broadcast mode, whereas the service announcement phase still exists. This allows potential service users to discover the availability of upcoming broadcast service sessions, together with relevant information about the service, its content and parameters required to receive the associated MBMS broadcast bearer.

The joining phase is also omitted, as the establishment of MBMS broadcast bearers is always network initiated and does not require any interaction with the service user. Service users interested in receiving a broadcast service only need to perform a local activation of the broadcast service, similarly to tuning in to an FM radio channel. The required information is either preconfigured or obtained via service announcements. The establishment of an MBMS broadcast bearer service is initiated by the BM-SC in the session start phase. As the interested service users are not known by the network, MBMS broadcast

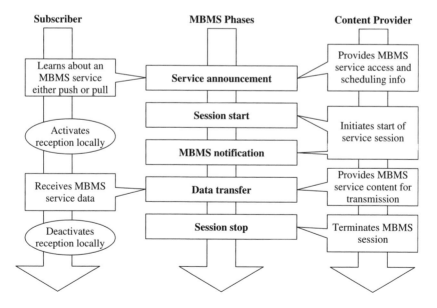

Figure 5.6 Service provisioning phases for MBMS broadcast services

bearers are established in all cells of the predefined broadcast service area. This can be less resource efficient than the establishment of the MBMS multicast bearer service, as in multicast mode the bearer service need only be established for cells in which multicast service users are present.

After the establishment of the broadcast bearers, data transfer can commence. As in multicast mode, MBMS notifications can be sent by the network in order to inform UEs about upcoming or ongoing broadcast service sessions. When the broadcast session is scheduled to end or no service data is available for transmission, the BM-SC can initiate the session stop phase. This results in the release of the MBMS broadcast bearer plane for the respective service.

Figure 5.7 illustrates a typical timeline for the different broadcast service provisioning phases. The first three horizontal lines from the top of the figure represent events, data transmission activity for the service announcement and data transmission for the MBMS service respectively. These three horizontal lines apply to an examplary MBMS broadcast user service A. The remaining horizontal lines 4 to 7 illustrate events and data transmission activity for two service users UE1 and UE2 respectively. UE1 and UE2 are assumed to be located in different cells. Following the event timeline of the broadcast service A, the service provider first schedules the service announcement for the upcoming

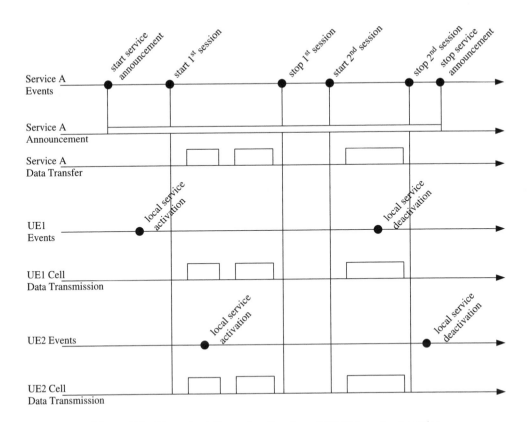

Figure 5.7   Example delivery timeline for a MBMS broadcast service

service sessions. The service announcement lasts for the entire duration of the two service sessions. The data transfer does not start directly after the session start of both service sessions but after a short idle period. This is important, as data should only be sent once the required broadcast bearer resources have been established. The first service session has an intermediate period of data inactivity in which the broadcast bearer resources are released by a session stop phase and resumed by a session start phase. In the second service session, data transfer is continuous for the entire duration of the session.

UE1 decides to receive the broadcast service after being informed of the service by the service announcement. UE1 activates reception for the broadcast service locally, without any interactions with the network. As advertised, the network transmits service data for broadcast service A in the cell in which UE1 is located. During the second session, UE2 decides to leave the broadcast service by deactivating the broadcast service locally. The network continues to broadcast the data until the end of the session, regardless of whether users are present in the cells of the service area or not.

UE2 also activates the reception of the broadcast service data after having received the service announcement. At the time of activation, the service session is already ongoing, and data is transmitted in the cell of UE2, regardless of whether there are any interested service users in the cell. UE2 leaves the broadcast service only after the end of the second session and therefore receives service data for the entire duration of the second session. As in the previous case, activation and deactivation of the broadcast service takes place locally at the receiver, without any involvement of the network.

5.3 BCMCS for CDMA2000

This section provides an overview of the main features of BCMCS. In many ways, BCMCS provides similar functionality to MBMS. As such, the section does not repeat what has already been described for MBMS in the previous section. Instead, this section focuses on explaining the commonalities and differences between BCMCS and MBMS. Many of the differences between BCMCS and MBMS are a result of the differences in the underlying network architectures of CDMA2000 and UMTS respectively. Also, different naming conventions are used in the specifications of both standards. A number of differences are down to design decisions that have been taken in the standardization bodies. As an example, the BCMCS standard does not support reliable download delivery required for file distribution and only provides support for best-effort delivery of real-time streaming traffic. MBMS, on the other hand, supports reliable data delivery.

In the following, an overview of the functional BCMCS network architecture is given, followed by a description of the core and access network extensions for BCMCS. Finally, the service provisioning phases for BCMCS are described.

5.3.1 Overview of BCMCS Architecture

The main objective of BCMCS is to support the efficient utilization of network resources in delivering content streams to multiple mobile subscribers. Besides functions for efficient transport, the BCMCS standard adds new functionality to the existing network architecture that allows the network operator to control the regions in which BCMCS is available to subscribers and accounting aspects for BCMCS, as well as the encryption of multicast flows in order to prevent unauthorized reception.

Unlike MBMS, BCMCS does not distinguish between a multicast and broadcast mode. Rather, the BCMCS standard distinguishes between static and dynamic broadcast. Dynamic broadcast is a broadcast service that establishes bearer paths in the network on the basis of the request of mobile subscribers. Static broadcast is a broadcast service in which bearer resources are provisioned statically by the operator, regardless of whether mobile subscribers are present. The BCMCS standard only describes procedures for the set-up of bearers for dynamic broadcast. The implementation of static broadcast is left as an implementation detail.

In BCMCS, services that make use of BCMCS capabilities are referred to as BCMCS broadcast programmes. A broadcast programme is usually identified by a programme name and a programme ID. Each programme provides a BCMCS content stream that consists of one or more IP multicast data flows. IP multicast flows are described by a destination IP multicast address, destination port number and source IP address and port number.

Figure 5.8 provides an overview of the BCMCS network architecture. The figure shows the existing logical network entities of a CDMA2000 network such as the MS, the BSC and PCF and several core network entities such as the PDSN and the network entity used for Authentication, Authorization and Accounting (AAA). In addition to these, the BCMCS standard introduces the Broadcast Serving Node (BSN), the BCMCS controller, the BCMCS content server and the BCMCS subscriber profile manager. Providers of BCMCS content streams can either be internal BCMCS content sources of the operator or third-party external BCMCS content providers.

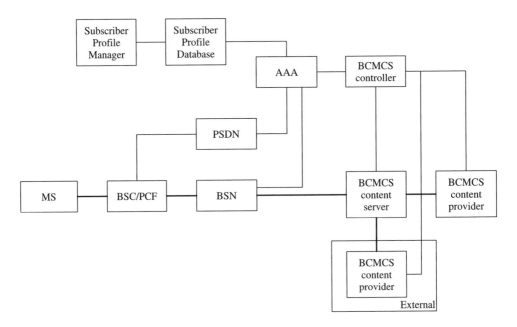

Figure 5.8   BCMCS reference architecture

5.3.2   Core Network Extensions

BCMCS introduces two logical entities, the BCMCS controller and the BCMCS content server, that provide the support functions for multicast and broadcast service provisioning. In terms of functionality, these entities can be compared with the BM-SC in MBMS. The BCMCS content server is the network entry point for broadcast programmes, making content available within the network as a BCMCS content stream. The BCMCS is not necessarily the source or creator of a broadcast programme, rather it stores and forwards content coming from internal or external content providers and may merge content from multiple content sources. The BCMCS is also able to manipulate the incoming content on the application level prior to being broadcast in the network, for example by transcoding to an appropriate bit rate.

While the BCMCS content server directly deals with the data plane, the BCMCS controller is responsible for managing BCMCS sessions in the network. The BCMCS typically provides session-related information to the BSN (via the AAA network entity), to the BCMCS content server and to the mobile users. BCMCS content providers usually interact with the BCMCS controller in order to configure session information such as the name of the content provider, the programme name, session descriptions and security information. Other session-related information exchanged between the entities include the session start time or the duration of the session. The BCMCS may perform authentication of BCMCS content providers. The BCMCS controller also interacts with the BCMCS content server to provide relevant session information, such as the scheduled start time or the expected duration of an BCMCS session. The BCMCS controller also allocates IP multicast addresses and port numbers for the different BCMCS control stream flows and manages the addition, modification or removals of BCMCS flows. The AAA network entity provides the BCMCS with authentication and authorization information of users as part of the BCMCS information acquisition procedure. The BCMCS controller may also send accounting information to the AAA and relay, via the AAA, session-related information to the BSN. The session-related information sent by the BCMCS controller contains details on the treatment of individual BCMCS flows, such as the IP multicast addresses and port numbers, BCMCS flow identifiers used in the RAN and QoS parameters such as delay or required bandwidth of these flows and whether or not header compression should be used. Besides the management of BCMCS-session-related information, the BCMCS controller is also responsible for the distribution of broadcast access keys. The BCMCS may also perform service discovery mechanisms to allow interested mobile subscribers to learn about available BCMCS broadcast programmes.

Another new logical entity added to the core network architecture is the BSN, which is located at the RAN edge and supports the provisioning of IP multicast flows to the RAN. The BSN usually terminates all IP multicast traffic from the core network and provides incoming BCMCS flows to the PCF for transmission over suitable radio bearers to the mobile users. The BSN handles BCMCS flows as indicated by the BCMCS controller. If a multicast-enabled backbone is utilized in the core network, the BSN uses IGMP or MLD to establish the necessary bearer path in the core network. This is done by subscribing to the multicast addresses of the corresponding flows at the closest multicast router within the backbone.

The final entity introduced into the core network architecture is the BCMCS subscriber profile manager. As with MBMS multicast services, mobile users may require a subscription for specific programmes or services. The subscriber profile manager allows either users or operators and their service providers to update and maintain a subscriber profile database. This database carries information on which BCMCS services a mobile user is subscribed to and thus authorized to receive.

5.3.3 Radio Access Network Extensions

The radio access network of CDMA2000 has been adapted in order to allow for efficient distribution of BCMCS content streams over the air interface. For this purpose, many of the existing physical channels defined by CDMA2000 have been reused. In this section, the key features of the high-rate broadcast-multicast packet data air interface for CDMA2000 EV-DO are presented. Further details on the high-rate broadcast-multicast packet data air interface can be found in (3GPP2 2006a).

The high-rate broadcast-multicast packet data air interface introduces a new broadcast channel, alongside a set of protocols. This set of protocols is referred to as the broadcast protocol suite. The broadcast protocol suite defines two subchannels to the broadcast channel, namely the basic broadcast channel and the enhanced broadcast channel. The *basic broadcast channel* is associated with a broadcast physical channel, which is similar to the forward traffic channel of the default physical-layer protocol of the CDMA2000 EV-DO standard. In contrast to the forward traffic channel, the broadcast physical channel supports soft-combining. The *enhanced broadcast channel* is associated with a broadcast physical channel that makes use of OFDM, a frequency-division multiplexing scheme also used in Digital Video Broadcasting (DVB) (DVB Project, 2008). The enhanced broadcast channel is thus not backwards compatible with the existing radio bearers. The structures of the different broadcast physical channels are defined by the MAC and physical-layer protocols of the broadcast protocol suite. Broadcast physical channels are subdivided into several time-division multiplexed subchannels, referred to as interlace-multiplex pairs.

BCMCS flows coming from the BSN are usually mapped onto broadcast logical channels. While a BMCMS flow is mapped only onto one specific broadcast logical channel, a single broadcast logical channel can carry more than one BCMCS flow. Broadcast logical channels are mapped to one or more interlace-multiplex pairs of the broadcast physical channel associated with a sector. As such, the broadcast logical channel is transported either on the basic broadcast channel or on the enhanced broadcast channel.

Figure 5.9 provides an overview of the broadcast protocol suite, which consists of the broadcast control protocol, the broadcast framing protocol, the broadcast MAC protocol and the broadcast physical-layer protocol. Details on the different protocols of the suite are given in the following.

Broadcast Control Protocol

The broadcast control protocol realizes the necessary control plane interactions between the user terminal and the RAN. It defines procedures for controlling various operational aspects of the broadcast channel. It enables user terminals to notify the RAN of their interest in receiving BCMCS flows, a process referred to as BCMCS flow registration. As can be seen in Figure 5.9, the broadcast control protocol makes use of the unicast MAC

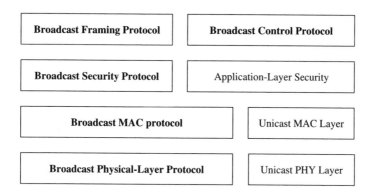

Figure 5.9 Broadcast protocol suite

and physical-layer protocols of the high-rate data packet air interface. This is necessary, as the signalling exchange of the broadcast control protocol is bidirectional and thus requires both downlink and uplink interactions.

Broadcast Framing Protocol

The broadcast framing protocol adapts higher-layer packets to the underlying frame characteristics of the broadcast channel. It typically performs fragmentation of high-layer SDUs on the RAN side, adding information necessary for determining higher-layer boundaries and integrity checking of received data frames. At the receiving side, the broadcast framing protocol defragments the received framing packets into higher-layer packets and validates the integrity of those by discarding erroneous packets. The broadcast framing protocol makes use of segment-based framing, instead of octet-based framing commonly used in framing protocols such as the High-Level Data Link Control (HDLC). Octet-based framing protocols require insertion of a flag sequence, resulting in higher deframing complexity, as the receiver needs to examine each octet of the received framing packet in order to determine the higher-layer packet boundaries. In addition, the flag sequence leads to a variable bandwidth expansion owing to insertion of escape sequences when higher-layer data matches the flag sequence. The segment-based framing used in the broadcast framing protocol makes use of an 8-bit frame header attached in front of each higher-layer payload fragment. Although several such headers may be present in a framing packet, the small size of 125 octets per MAC-layer packet makes it highly unlikely that more than two such frame headers occur in a single framing packet. Instead of examining each octet only, the framing headers need to be examined, resulting in a much lower processing complexity at the receiver for the defragmentation processes. Also, bandwidth expansion is up to 7 times lower compared with HDLC-like framing (Agashe, Rezaiifar and Bendar 2004).

Broadcast Security Protocol

In order to control the access to broadcast content, encryption can be used as a useful security mechanism. Encryption of broadcast content before transmission allows only receivers with a suitable decryption key to gain access to the content after transmission. The BCMCS standard defines two main aspects of broadcast security. Key management handles the secure distribution of decryption keys for BCMCS sessions to authorized subscribers, whereas the encryption of broadcast content provides the required security. In this section, we only describe the mechanisms for encryption. Details on key management in the broadcast security framework can be found in (3GPP2, 2003).

Unlike in MBMS, the encryption and decryption of broadcast content in BCMCS can take place within the RAN and are realized by the broadcast security protocol. If encryption is desired, the broadcast security protocol on the access network side will encrypt the framing packets generated by the framing protocol and pass the so-called generated security packets on to the MAC protocol for further processing prior to transmission. If no encryption is desired, the broadcast protocol will leave the framing packets untouched. The broadcast security protocol makes use of the AES encryption procedures (3GPP2, 2008), using a Short-Term Key (SK) that is also known on the receiver side for the decryption process. The SK is only valid for a limited amount of time, in most cases a single broadcast session, and is also never transmitted over the air. Instead, it is derived at the receiver from a Broadcast Access Key (BAK), associated with a particular BCMCS programme, and a random value broadcast along with the receiving content. Using the SK and a cryptosync, the receiver generates an encryption mask. The framed packets are then XOR-ed with corresponding portions of the encryption mask to generate the desired security packet. At the receiver side, the broadcast security protocol generates the necessary decryption keys from the local BAK and the transmitted random seeds. It then decrypts the security packets before passing them on as framing packets to the framing protocol.

Broadcast MAC Protocols

Error control in the CDMA2000 EV-DO standard relies on ARQ. As no reverse link is available on the broadcast channel to transport acknowlegements to the RAN MAC entities, an FEC-based error control scheme is used for the broadcast MAC protocol. Using a Reed-Solomon block code, the broadcast MAC protocol adds an outer code to the security packets received by the broadcast security protocol. This forms a product code in conjunction with the turbo code of the physical layer. When used as outer codes, Reed-Solomon codes perform well at lower error rates, requiring less decoding complexity than convolutional codes (Agashe, Rezaiifar and Bender, 2004). At the receiver side, the Reed-Solomon code is used as an erasure code, relying on the CRC of the physical layer to detect damaged physical-layer packets. This simplifies the recovery of erased octets by the Reed-Solomon code, as the position of erased octets is known in advance. More details on the broadcast MAC protocol can be found in (Agashe, Rezaiifar and Bender, 2004) and (3GPP2, 2007a).

Broadcast Physical-Layer Protocols

The broadcast physical-layer protocol introduces soft-combining on the forward link during broadcast service transmission. As in the MBMS, soft-combining is used as a form of macrodiversity to improve the block error rate of the transmission. This is done at physical layer by combining the signals received from multiple adjacent cells or sectors that transmit the same data. More details on the broadcast physical-layer protocol can be found in (Agashe, Rezaiifar and Bender, 2004) and (3GPP2, 2007a).

5.3.4   Service Provisioning Phases

The service provisioning phases in BCMCS are conceptually similar to the service provisioning phases in MBMS. In spite of several differences in terminology, several service provisioning phases in BCMCS can be mapped one-to-one to a corresponding service provisioning phase in MBMS. The procedures executed during these phases can be slightly different owing to architectural difference between BCMCS and MBMS. In the following, the BCMCS service provisioning phases for the BCMCS dynamic broadcast service are examined.

As mentioned earlier, the BCMCS dynamic broadcast service requires interested users to subscribe to a particular BCMCS programme. For static broadcast services, this step is not required. Users may choose to subscribe to one or more BCMCS programmes according to their interest; this information is recorded by the BCMCS subscriber profile manager. The mechanism for subscription is not specified by the BCMCS standard. Information about services can be obtained either offline or by service announcements.

The service provisioning phases for BCMCS are illustrated in Figure 5.10. A user typically learns about the existence of BCMCS programmes via some form of service announcement that is performed by the operator or a BCMCS programme provider during

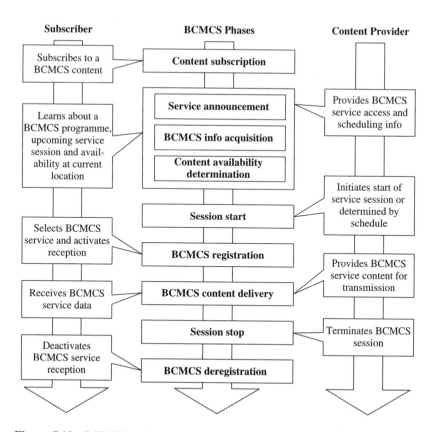

Figure 5.10   BCMCS service provisioning phases for dynamic broadcast services

the service announcement phase. The service announcement can be realized via different mechanisms such as SMS or WAP. In BCMCS, the service announcement phase is a form of initial high-level service discovery. Service announcements usually carry information about the service, such as a programme description, and the parameters required for information acquisition, such as the programme name or programme identifier. Service announcements may also carry other service-related information such as the start time of the service. Unlike in MBMS, the user terminal is not provided with sufficient information to join one or more BCMCS service flows of an upcoming session.

The information required to do so is obtained in a separate step referred to as the BCMCS information acquisition. During the BCMCS information acquisition phase, the user interacts with the BCMCS controller via the HTTP protocol. The user usually provides some identifier of a particular service obtained during the service announcement to obtain the required information in order to receive one or more BCMCS flows of a service session. The user can also request information for multiple programmes at the same time, or even for all available services by not providing any identifier at all. For each requested programme, the BCMCS controller then returns several parameters in response to the BCMCS information acquisition request, including the programme name and description and the schedule time, as well as information for each BCMCS flow of the session. The information for each BCMCS flow may consist of the flow description, the multicast IP address and port numbers, the transport protocol to be used and application-layer codecs. The BCMCS controller also provides link-layer information such as details on the encryption methods and header compression, as well as several security-related parameters.

After the BCMCS information acquisition, the user terminal has acquired all information necessary to receive the desired BCMCS flows of a service session. However, the user terminal is not aware of whether these service flows are actually offered in the cell in which the user is located. During the so-called content availability determination phase, the user terminal discovers whether the service is available and obtains necessary BCMCS radio configuration parameters from the base station via CDMA2000 overhead messages. In case the user terminal does not find any overhead messages even though support for the service has been indicated (for example, if it is the first user terminal in the cell requesting a particular BCMCS flow), it can request the BCMCS radio configuration parameters during the BCMCS registration phase.

The BCMCS registration phase is similar to the joining phase in MBMS. In this phase, a user can notify the RAN of the BCMCS programmes and service flows in which it is interested. Unlike in MBMS, the user terminal in BCMCS does not make use of IGMP or MLD messages for registration, but instead uses BCMCS-specific signalling mechanisms to inform the RAN. The network may require authorization of the BCMCS registration message in order to ensure that only authorized user terminals can gain access. The network may also require user terminals periodically to re-register, allowing the network to monitor the continuing interest for a given BCMCS flow.

The first registration of a user terminal for a BCMCS flow triggers the RAN session discovery procedure. With this procedure, the RAN is able to obtain session-related parameters for the BCMCS flow, such as the session start time and duration. The RAN session discovery is a prerequisite for the set-up of BCMCS bearers. The BCMCS registration may thus result in the establishment of core network and RAN bearers for the BCMCS flow, given that the bearer path is not provisioned statically. Bearer path establishment can also be triggered by the BCMCS controller.

The BCMCS content delivery phase is equivalent to the MBMS data transfer phase. In this phase, BCMCS session flows are transmitted via the established bearer paths to the receivers. The RAN may apply encryption and header compression before sending the session data over the air interface. During this phase, late-arriving users can also register for BCMCS flows. BCMCS deregistration corresponds to the leaving phase in MBMS. Here, the network determines whether bearer resources for BCMCS flows are still required. BCMCS deregistration can be performed in two possible ways. With the first mechanism, the user terminal explicitly sends BCMCS deregistration signalling to the RAN as soon as a user decides to stop receiving a particular BCMCS programme. With the second mechanism, the operator requires user terminals periodically to re-register for the BCMCS flows they are currently receiving and can implicitly assume deregistration after expiration of a time-out. Finally, on termination of a BCMCS session, the network may release the BCMCS bearer path. Furthermore, the network can indicate to user terminals that a session has stopped and registration for the BCMCS session is not allowed any more.

With static broadcast in BCMCS, the service delivery phases are simpler, similar to the service delivery phases in MBMS broadcast mode. As an example, content subscription in most cases is optional. All bearer paths in the network are statically provisioned by appropriate out-of-band mechanisms not specified in the BCMCS standard. As such, BCMCS registration and RAN session discovery phases are no longer required. Likewise, BCMCS deregistration is also not required.

5.4 Summary

This chapter has provided an initial overview of the architectural extensions required to support the efficient delivery of multicast and broadcast services in UMTS and CDMA2000 networks. The MBMS standard provides multicast capabilities in UMTS networks, whereas the BCMCS standard provides multicast capabilities in CDMA2000 networks. Both MBMS and BCMCS can be used to deliver content to a potentially large number of receivers. The introduction of MBMS and BCMCS has resulted in changes to the core and radio access networks of UMTS and CDMA2000 respectively.

For MBMS, the BM-SC network node has been added to the UMTS core network. The BM-SC is the entry point for content providers that wish to offer MBMS services. Also, functionality for the management of MBMS bearers has been added to the GGSN and SGSN network nodes. In the radio access network, three new logical channels have been introduced for point-to-multipoint transmissions. All three channels are multiplexed onto the existing FACH in UTRAN. Also, the MAC layer on the RAN and UE side has been adapted for the purposes of MBMS. As in the core network, functionality for the management of MBMS radio bearers has been added to the radio access network.

The introduction of BCMCS in CDMA2000 has made changes to both core network and radio access network necessary. The main new logical entities are the BCMCS controller and the BCMCS content server. The BCMCS content server is the network entry point for broadcast programmes, making any content available in the form of BCMCS content streams. While the BCMCS content server is responsible for the data plane, the BCMCS controller is responsible for managing the BCMCS sessions on the control plane. The combination of the BCMCS content server and the BCMCS controller is comparable with the functionality of the BC-SC. Two physical channels, the basic broadcast and enhanced broadcast channels, can be used in BCMCS in order to provide efficient downlink transmission of BCMCS programmes over the air interface. The basic broadcast channel is compatible with the existing unicast high-rate data packet air interface of CDMA2000, whereas the enhanced broadcast channel is not, as it makes use of OFDM.

In many cases, the functionality of MBMS and BCMCS is very similar. Many differences can be attributed to the different system architectures of the underlying mobile networks. One difference between MBMS and BCMCS worth mentioning is that BCMCS is predominantly designed for streaming-type services, whereas MBMS supports both streaming-type and file delivery services. BCMCS is not suited for file delivery services, as absolute delivery guarantees cannot be provided on an end-to-end basis. In contrast, the MBMS standard provides end-to-end reliability between the sender and all receivers using FEC-based error protection and an optional file repair service at the end of a session to recover from losses.

6

Protocols and Mechanisms for MBMS

6.1 Introduction

The previous chapter provided an overview of multicast mechanisms in UMTS and CDMA2000 networks. This chapter focuses on MBMS, the multicast and broadcast standard for UMTS. Besides the introduction of multicast and broadcast bearer capabilities, MBMS introduces a complete service architecture together with a framework for the provision of MBMS services. This service architecture and framework, depicted in Figure 6.1, defines a set of service functions that may be used for the efficient provision of MBMS user services in UMTS networks.

The MBMS service architecture can roughly be split into two layers. The service layer encompasses user service functions that are necessary for the set-up and management of MBMS service sessions and provides application-specific support for the delivery of content, whereas the network layer provides bearer services for the transport of service data, related control signalling and service management information. Architecturally, the service layer consists of functions in the BM-SC and the application entities of the UE. The network layer consists of functions provided by the other UMTS network entities.

Within the MBMS service architecture, multicast and broadcast services are provided as MBMS user services. The service layer offers two delivery methods, namely streaming and download delivery. The streaming delivery method provides a set of protocols and codecs suitable for the delivery of streaming sessions such as real-time audio or video feeds. The download delivery method provides a set of protocols and support mechanisms for the reliable distribution of data files and software, which is typically not delay sensitive. The service layer also provides support mechanisms for the management of user service sessions such as service discovery and access management in the form of security-related functions.

In order to deliver the user service data, the user service makes use of one or more bearer services provided by the network layer. As mentioned earlier, the MBMS standard

Multicast in Third-Generation Mobile Networks   Robert Rümmler, Alexander Gluhak and A. Hamid Aghvami
© 2009 John Wiley & Sons, Ltd

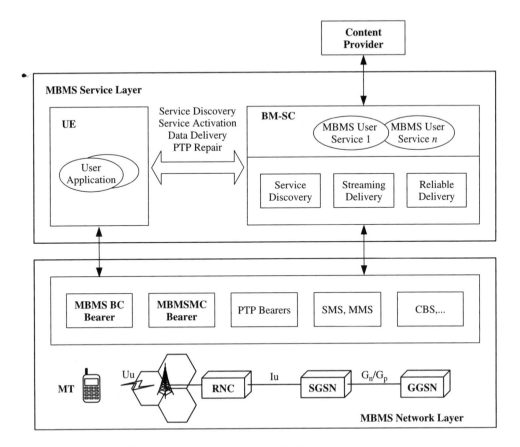

Figure 6.1   Overview of the MBMS service architecture

introduces broadcast and multicast bearer services, which are mainly used for the transport of user service data. Other bearer capabilities such as point-to-point bearers or alternative delivery methods such as SMS, MMS or CBS can be used for other service-related interactions such as service discovery, point-to-point file repair or user registrations. At the UE, the bearer services are terminated at the Mobile Terminal (MT). At the MT, the bearer services are finally accessed by application entities.

The rest of the chapter is organized as follows. Firstly, details of the MBMS multicast and broadcast bearer services are presented. This includes basic concepts such as the associated bearer contexts and the supported QoS as well as management mechanisms of the MBMS bearer plane. Next, an overview of routing along the bearer path is provided, exploring the different steps involved in routing packets towards the group of receivers. The final part of the chapter deals with the service layer of the MBMS service architecture. User services and their interactions with MBMS bearers are discussed in detail. Also, the protocols and mechanisms used for the different delivery methods and support services are described. The chapter closes with a summary and some closing remarks.

6.2 MBMS Bearer Service Basics

This section introduces the basics of MBMS bearer services by introducing the MBMS bearer service architecture, the MBMS bearer context and the MBMS UE context.

6.2.1 MBMS Bearer Service Architecture

UMTS networks are required to support the transport of user data for a variety of different applications. The requirements of these applications on the underlying transport service vary widely and are often driven by the perceived QoS of a user. Audio and video streaming, for example, require ordered delivery with low delivery latency but can tolerate a small amount of data loss, while the delivery of files is usually not delay sensitive but requires reliable delivery without the loss of information. In order to realize data transport with different QoS attributes, the UMTS architecture defines the concept of bearers (3GPP, 2007h). A bearer is an information transmission path with defined QoS attributes such as capacity, delay, bit error rate and so forth. A bearer service includes all aspects that enable the transmission of signals at a required QoS between two communication endpoints, such as control signalling, user plane transport and QoS management functionality.

Figure 6.2 provides an overview of the UTMS bearer service architecture. As can be seen from the figure, the bearer services are offered by different layers, and bearer services on a higher layer make use of the services offered by the bearer services below them. The transmission of data between two application endpoints is realized by the end-to-end service, which makes use of the underlying bearer services provided by the different network nodes along the path. Any bearers outside the UMTS network are referred to as external bearer services, while bearers inside the UMTS network make use of UMTS bearer services. As the bearer service terminates in the MT, a local bearer service complements the UMTS bearer service to form an end-to-end service between the application entities. For reasons of simplicitiy, this is omitted in Figure 6.2.

The transmission path within the UTMS network may consist of different communication technologies. These communication technologies usually provide varying transmission characteristics and therefore require slightly different QoS control mechanisms to achieve the desired QoS characteristics. IP-based network technology is often used between the network elements of the core network. Radio access network connectivity is often realized via ATM, while radio communication is realized over the air interface.

The UMTS network bearer service is realized by a core network bearer service between the network elements of the core network and a radio access bearer service from the edge of the core network to the UE. The core network bearer service in turn controls the backbone network, which provides backbone bearer services supporting a variety of QoS. The radio access bearer service is realized by the radio bearer service and a RAN access bearer service, which in turn makes use and controls the underlying physical bearer services.

The QoS management across different bearer services is realized by QoS management functions, located at the different communication endpoints in the UMTS network entities and at different layers. Control plane management functions allow the establishment and modification of bearer services with required QoS attributes at the different layers within

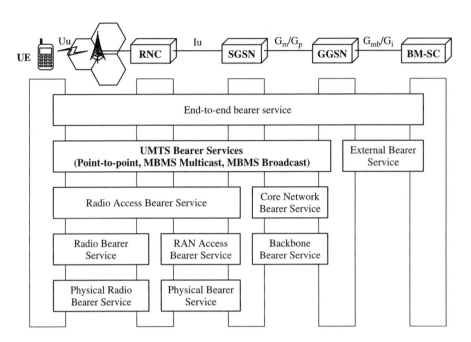

Figure 6.2 UMTS bearer service architecture

the UMTS network. Control plane management also allows for a mapping to and from external bearer services to ensure adequate QoS support for the end-to-end service. User plane management functions ensure that the data transfer takes place according to the committed bearer service attributes established by the control plane, using functions such as traffic conditioning to adapt the data flows to the underlying bearer characteristics.

For point-to-point bearer services, the UMTS architecture defines four QoS classes, namely the conversational class, the streaming class, the interactive class and the background class. The main differentiating characteristic between the QoS classes, which are often referred to as traffic classes, is the delay sensitivity of the traffic. The conversational class provides the lowest delivery delay, while the background class, located at the other end of the spectrum, assumes traffic to be agnostic to delay. Both the conversational and the streaming class are intended for real-time traffic and aim at keeping the time variations or jitter between information elements within a stream as small as possible. The streaming class is typically used for streaming audio and video content to mobile users. The conversational class imposes very stringent requirements on the delivery delay and is suitable for highly delay-sensitive traffic such as voice over telephony, voice over IP or video conferencing. In contrast, the interactive and the background classes are more suited towards delay-insensitive traffic and for traffic which requires reliable delivery. The background class is useful for applications with no real-time delivery requirements, such as the background download of email or software updates. The interactive class provides higher responsiveness and is suited to interactive applications such as web browsing, where the user expects a response to his request in a reasonable timeframe. This is achieved by prioritizing the transmission of interactive class traffic compared with background class traffic. For point-to-multipoint delivery of data via MBMS broadcast and multicast bearer services, only two of the four traffic classes are supported, namely the streaming and the background class. A detailed discussion of the different traffic classes and corresponding bearer service attributes is provided in (3GPP, 2007h).

The difference between MBMS bearer services of the background and the streaming class is that the latter provides support for a guaranteed bit rate. MBMS bearer services of the streaming class are most suited for the transport of MBMS user services that provide some form of real-time content such as audio or video streams. The QoS management mechanisms in the network ensure that the transfer delay of packets along the bearer path is minimized. Dropping packets in case of congestion is the preferred traffic conditioning method applied to the traffic flow. MBMS bearer services of the background class are most used for the download delivery of data files or messages. In order to adapt the traffic flows of the background class, the QoS management mechanism can buffer packets at intermediate nodes along the path or choose to drop packets completely. This usually requires MBMS user services to provide higher redundancy in error coding or, in the case of an unacceptable QoS, a change to the streaming bearer service. The lack of a feedback channel over the radio bearer path does not allow low SDU error rates to be achieved for the background class. In this case, the user services may offer a file repair service by transmitting missing packets over point-to-point bearer services to individual receivers.

As MBMS bearer services rely on point-to-multipoint distribution trees, it may not be possible to guarantee the same QoS on all branches of the distribution tree. MBMS requires the establishment of bearer paths that provide homogeneous QoS support across the whole distribution tree. The addition of a branch with a different QoS is not permitted when a branch of the MBMS distribution tree at a specific QoS already exists. As a result, some branches may not be established and the respective receivers under this branch are left without support for the MBMS session.

6.2.2   MBMS Bearer Context

Each network node along a bearer path maintains the relevant state for an MBMS bearer service in an MBMS bearer context. The MBMS bearer context usually contains all state information required for packet forwarding and the treatment of service data and control signalling for the associated MBMS bearer service. Within the RAN, the MBMS bearer context is also referred to as an MBMS service context.

MBMS bearer contexts in different network nodes are different, as shown in Figure 6.3. Common information elements in the MBMS bearer contexts of all network nodes are the Broadcast/Multicast (BC/MC) mode indicator, status, QoS attributes, Temporary Mobile Group Identity (TMGI) and the MBMS service area. The BC/MC mode indicator defines whether the MBMS bearer service operates in multicast or broadcast mode. In the case of multicast mode, each MBMS bearer context would contain a corresponding IP multicast address and an associated Access Point Name (APN), in which the scope of the IP multicast address is defined. The IP multicast address is used by the UEs to join a particular MBMS bearer service. The status field indicates whether or not bearer plane resources are required for an imminent data transfer. The QoS attributes define the manner in which service data is treated by QoS management mechanisms. The TMGI is a temporary identifier assigned by the BM-SC, which identifies an MBMS bearer service. The MBMS service area element defines the area of the UMTS network in which data of an MBMS

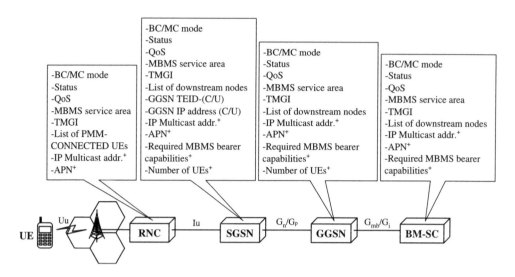

Figure 6.3   MBMS bearer contexts in the different nodes of the network

bearer service needs to be distributed. MBMS bearer context information elements that are specific to multicast mode are labelled with a plus sign in the figure.

The MBMS bearer service forms an efficient distribution tree from the BM-SC down to the radio access networks in which interested receivers are located. This tree is established by bearer management procedures described in the following section, and is realized by keeping a list of downstream nodes in each network node. Each network node in the delivery tree receives service data for an MBMS bearer service from its upstream node and forwards the service data only to its downstream nodes that are included as an entry in the list. The SGSN always maintains the IP addresses and TEID of the GGSNs involved in the MBMS service, for both the user and control plane. Both the SGSN and the GGSN keep track of the number of UEs that have joined the MBMS multicast service. The RAN maintains a similar list, but only for PMM-CONNECTED UEs.

6.2.3 MBMS UE Context

Besides MBMS bearer contexts that contain bearer-specific information for an MBMS bearer service, the UMTS network also maintains user-specific information for MBMS bearer services in multicast mode. This allows individual users to be billed. This user-specific information is referred to as an MBMS UE context and exists for UEs that have activated an MBMS multicast bearer service. The MBMS UE context is stored as part of the MM context of an UE. For each UE, there is typically one MBMS UE context per MBMS multicast bearer service a UE has joined. The MBMS UE contexts for a particular MBMS bearer service are usually linked to the MBMS bearer context of that MBMS bearer service.

Figure 6.4 provides an overview of the different information elements that are kept as part of an MBMS UE context in the different nodes of the network. The information elements of the MBMS UE context that are common in all network nodes are the IP multicast address, APN and the International Mobile Subscriber Identity (IMSI). The

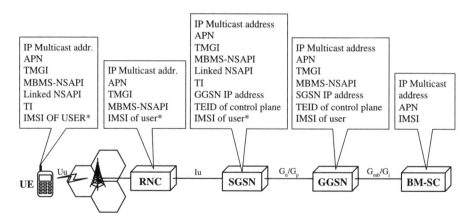

Figure 6.4   MBMS user context in the network nodes

multicast address and APN serve as unique identifiers of the MBMS bearer service, whereas the IMSI serves as a unique identifier of the subscriber. The IMSI is stored as part of the MM context in the UE and the SGSN, while the IMSI is stored as part of the UE context in the RAN. Except in the BM-SC, the TMGI, an identifier of an MBMS bearer service that is efficient in terms of radio resources, is present in the MBMS UE contexts of all network nodes. An MBMS UE context is identified by an MBMS Network Layer Service Access Point Identifier (NSAPI). In order to activate or deactivate an MBMS multicast bearer service, the UE typically joins or leaves an IP multicast group by a PDP context, identified by an NSAPI. Both the UE and the SGSN maintain a link to the NSAPI of the unicast PDP context as part of the MBMS UE context as well as a corresponding transaction identifier. For UE-specific control signalling, both the SGSN and GGSN maintain IP address information and the TEID of the corresponding control plane. Also, additional information for operational management can be maintained as part of the MBMS UE context.

6.3   MBMS Bearer Service Management

The management of MBMS bearer services is governed by a set of MBMS procedures. These MBMS procedures specify how and when MBMS bearer services and the associated bearer state are created or released within the UMTS network and how the bearer service paths are managed during service provisioning.

The establishment and release of MBMS multicast bearer services is receiver initiated and triggered by the MBMS multicast service activation and deactivation procedures. The MBMS UE linking and delinking procedure are used to provide MBMS UE context to the UTRAN. The MBMS registration and deregistration procedures are used to maintain an accurate delivery path for MBMS bearer services. The MBMS session start and session stop procedures are used to allocate and release bearer plane resources for data transfer along the bearer path. The MBMS update procedure is used to update bearer service attributes such as the MBMS service area for an already established bearer service. In order to manage the bearer services in case of user mobility, existing UMTS mobility procedures, namely the inter-SGSN routing area update procedure and the inter-SGSN serving RNS relocation procedures, have been extended to support MBMS bearer services. Finally, the MBMS notification procedure is used at the start of a session to inform UEs of an upcoming MBMS session. The MBMS service request procedure allows the RAN to determine the approximate number of MBMS users for an MBMS multicast service. Based on this information, the RAN is able to select for each cell whether PTP or PTM radio bearers should be established.

Many of the MBMS procedures require interactions across multiple reference points of the UMTS architecture and involve different protocol entities. Existing protocols such as GTP-C or Radio Access Network Application Part (RANAP) have been enhanced to accommodate the MBMS procedures. The remainder of this section describes these procedures in more detail.

6.3.1 MBMS Activation and Deactivation

The MBMS multicast service activation procedure is initiated by a UE to register its interest in receiving MBMS user service data. The procedure can initiate the establishment of an MBMS multicast bearer path (if not yet previously triggered by another UE in the same area) and establishes MBMS UE contexts in the UE, SGSN, GGSN and UTRAN for the activated MBMS bearer service.

Figure 6.5 provides an overview of the signalling flow of the MBMS multicast service activation procedure. The activation is typically performed via a default best-effort unicast PDP context, which the interested receiver has to establish for group management signalling using IGMP or MLD. In the following, each step in Figure 6.5 is explained in more detail, with RES and RES in the figure being short for request and response respectively.

1. The UE sends an IGMP or MLD membership report via the pre-established default PDP context to the GGSN, identifying the IP multicast address of the MBMS bearer service it is interested to join. The IP multicast address is known from an MBMS service announcement.
2. The GGSN intercepts the IGMP or MLD membership report, extracts the IP multicast address and (if received for the first time for that UE) tries to obtain MBMS authorization from the BM-SC for the activating UE. The group membership function in the BM-SC may check whether the user has a valid subscription. If so, it provides the APN to be used for the creation of the UE context in the MBMS authorization response. The procedure terminates if authorization fails.
3. The GGSN sends an MBMS notification request providing the SGSN with the IP multicast address, APN and linked NSAPI of the PDP context on which the IGMP signalling was received. In the MBMS notification response the SGSN then indicates to the GGSN whether the procedure is able to proceed, that is, if both the UE and SGSN support MBMS.
4. The SGSN then initiates the activation of an MBMS UE context at the UE, providing the IP multicast address, APN, linked NSAPI and a unique transaction identifier with the request MBMS context activation message.
5. The UE creates an MBMS UE context with the parameters provided by the SGSN and sends an activate MBMS context request to the SGSN, providing all previous

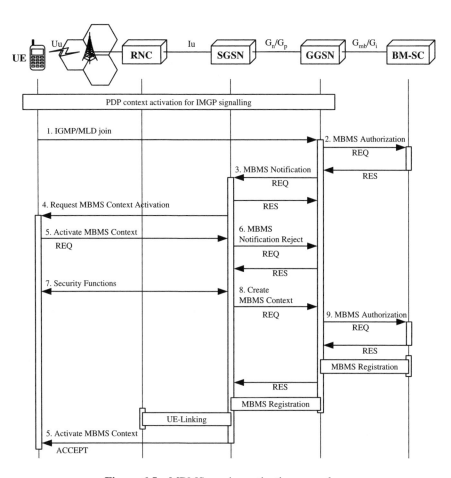

Figure 6.5 MBMS service activation procedure

parameters to identify the UE context including its own MBMS bearer capabilities, indicating the maximum QoS it can handle. If an MBMS bearer context already exists in the SGSN, then the SGSN verifies that the UE-supported MBMS bearer capabilities are not lower than required for the MBMS session. If so, the activate MBMS context request for the UE is rejected. Otherwise, the procedure below continues and a successful termination is indicated to the UE via the activate MBMS context accept message.

6. If for some reason the MBMS UE context was not established, the SGSN notifies the GGSN about this failure. This is done by indicating the failure via the MBMS notification reject request message. The GGSN then deletes all related state and indicates the termination of the procedure to the SGSN with the MBMS notification reject response.

7. Security functions such as the authentication of a UE may be performed.

8. The SGSN creates an MBMS UE context and sends a *create MBMS context request* to the GGSN, identified by the associated APN of the multicast bearer service. Besides the IP multicast address, the APN and the MBMS NSAPI, the request contains a variety of different UE-related parameters such as the IMSI, the Mobile Subscriber ISDN Number (MSISDN), the telephone number of the subscriber or the IMEI.

9. The GGSN then tries to obtain authorization for the UE by sending an MBMS authorization request to the BM-SC. After successful authorization, the BM-SC creates an
MBMS UE context for the user and informs the GGSN with an MBMS authorization
response about the outcome of the authorization. After successful authorization, the
GGSN creates an MBMS UE context and sends a *create MBMS context response* to
the SGSN. If the GGSN does not have an MBMS bearer context for the activated service, it triggers the execution of the MBMS registration procedure, which is explained
in the following section. Likewise, the SGSN triggers the execution of the MBMS
registration procedure if the MBMS bearer context for the activated service does not
yet exist. The SGSN also provides the UE context to the RAN via the MBMS UE
linking procedure, if at least one packet-switched RAB exists for the UE.

MBMS multicast bearer services can also be activated without the use of IGMP or
MLD signalling for bearer service control. In this case, the UE sends an *activate MBMS
context request* directly to GGSN, thus omitting Steps 1 to 4 above.

Figure 6.6 depicts the MBMS multicast service deactivation procedure that removes the
MBMS UE context from the UE and the other network elements. Unlike the activation
procedure, which must be initiated by the UE, the deactivation procedure can be initiated
by the UE, GGSN, SGSN or BM-SC. The UE-initiated deactivation procedure starts at
Step 1 or alternatively at Step 6 if the deactivation procedure is not based on IGMLP or
MLD. The deactivation procedure initiated by the BM-SC starts at Step 3, whereas the
deactivation procedure initiated by the GGSN starts at Step 4. The deactivation procedure
initiated by the SGSN starts at Step 5. The MBMS deactivation procedure depicted in
Figure 6.6 is described as follows:

1. The UE sends an IGMP or MLD membership report via the default PDP context,
   indicating the IP multicast address of the MBMS multicast service it would like to
   leave.
2. The GGSN receives the leave message and sends a leave indication to the BM-SC,
   providing the IP multicast address, APN and IMSI of the UE.
3. The BM-SC verifies whether an MBMS bearer service exists with the indicated IP
   multicast address and sends a UE removal request to the GGSN, providing the IP
   multicast address and the APN as an identifier of the multicast bearer service and the
   IMSI as an identity of the UE.
4. The GGSN sends an MBMS UE deactivation request to the SGSN, providing the IP
   multicast address, the APN as well as the IMSI. The SGSN acknowledges the message
   by sending an MBMS UE deactivation response back to the GGSN.
5. The SGSN sends an MBMS deactivate UE context request to initiate the deletion of
   the MBMS UE context at the UE. Based on the TI provided with the message, the
   UE is able to identify the necessary MBMS UE context for deletion. The successful
   deletion of the MBMS UE context is confirmed by the MBMS deactivate UE context
   response. The deletion of the MBMS UE context can also be initiated by the UE if
   signalling based on IGMP or MLD is not used. In this case, the UE sends the MBMS
   deactivate UE context request instead.
6. If the UE is linked to the RAN, the SGSN sends an MBMS UE delinking request to
   the RAN, providing the IP address, APN and TMGI. The RNC deletes the MBMS UE
   context and confirms this by sending an MBMS UE delinking response.

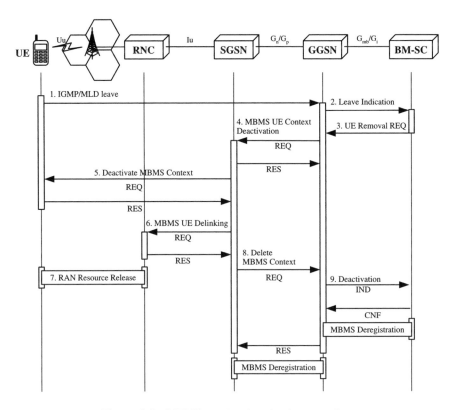

Figure 6.6   MBMS service deactivation procedure

7. The RAN may choose to evaluate the continued use of existing bearer resources and may release dedicated or shared bearer resources that have been in use for the UE.
8. The SGSN then initiates the deletion of the MBME UE context at the GGSN by sending a *delete MBMS context* request to the GGSN that holds the MBMS UE context. The respective UE context is identified by an MBMS NSAPI.
9. The GGSN deletes the MBMS UE context and sends a deactivation indication to the BM-SC to confirm the successful deactivation. The BM-SC also deletes the MBMS UE context and sends a confirmation back to the GGSN. If the GGSN has no more interested users for a particular MBMS multicast service and the list of downlink nodes attribute of the MBMS bearer context is empty, the GGSN performs an MBMS deregistration. The MBMS deregistration procedure is described in the following section. The GGSN confirms the deactivation of the MBMS UE context to the SGSN, with the SGSN also evaluating the number of users and the list of downlink nodes in the MBMS bearer context and potentially triggering the MBMS deregistration procedure to update the delivery tree.

6.3.2   MBMS Registration and Deregistration

The MBMS registration and deregistration procedures are used for the management of the MBMS multicast distribution tree within the UMTS network. The registration procedure is used by a downlink node such as the RNC, SGSN or GGSN to inform its uplink node of interest in receiving session signalling, session attributes and data for a particular MBMS bearer service. The requested uplink node typically adds the requesting downlink node into its list of downlink nodes attribute of the corresponding MBMS bearer context and provides the downlink node with the information necessary for the construction of an MBMS bearer context. The MBMS registration procedure results in the creation of MBMS bearer context in the requesting node.

There are several ways in which the MBMS registration procedure can be initiated. It can be invoked by a network node when the first MBMS UE context is created for an MBMS bearer service and the corresponding bearer context is not available at the node. It can also be triggered by the reception of an MBMS registration request from a downstream node for an MBMS bearer service for which the node does not have the corresponding bearer context. Finally, the MBMS procedure can also be invoked by an RNC that detects that it is hosting UEs interested in a particular MBMS bearer service, but that it does not have an MBMS bearer context for the MBMS bearer service in question. Figure 6.7 illustrates the signalling steps of the MBMS registration procedure. The signalling flow is described in the following:

1. An RNC sends an MBMS registration request to its parent SGSN if it is informed, for example by the MBMS UE linking procedure, that it is serving a UE that is interested in an MBMS bearer service and does not hold a corresponding MBMS bearer context. It provides the IP multicast address and the APN to identify the desired MBMS bearer service. The SGSN verifies whether it holds a corresponding MBMS bearer context. If the SGSN already has the MBMS bearer context, it simply adds the RNC address to the list of downstream nodes attribute and provides relevant bearer context information such as the TMGI and the bearer capabilities in the MBMS registration response. Based on this information, the RNC creates the corresponding MBMS bearer context

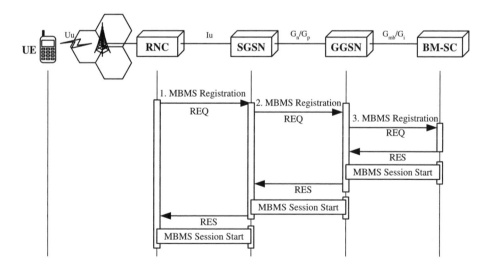

Figure 6.7 MBMS service registration procedure

and sets the state variable to inactive. If the bearer context state at the SGSN is active, the SGSN also triggers the session start procedure.

2. If the MBMS bearer context is not available at the SGSN, the SGSN creates an MBMS bearer context, setting its status attribute to standby and adding the identifier of the requesting RNC to the list of downlink nodes attribute of the newly created context. It then requests bearer context information by sending an MBMS registration message to the GGSN that is responsible for the MBMS bearer service. This is done by resolving the APN. Based on the APN and the IP multicast address information provided in the MBMS registration request, the GGSN searches for a matching MBMS bearer context. If it finds a match, the GGSN adds the identifier of the requesting SGSN into the list of downstream nodes attribute and provides the relevant bearer context information in the MBMS registration response. The SGSN then completes the MBMS bearer context with this information and responds to its respective downstream nodes with outstanding MBMS registration requests.

3. If the GGSN is unable to find the MBMS bearer context for a MBMS bearer service, it creates a new MBMS bearer context, setting its status attribute to standby and adding the SGSN identifier to the list of downlink nodes attribute of the newly created context. The GGSN then sends a registration request message to the BM-SC, providing the IP multicast address and the APN to identify the MBMS bearer service. The BM-SC adds the identifier of the GGSN into the list of downlink nodes attribute of the corresponding MBMS bearer context and provides bearer context information such as the TMGI as well as bearer capabilities in the MBMS registration response. The GGSN then completes the MBMS bearer context and responds to its respective downstream nodes with outstanding MBMS registration requests. If the MBMS bearer context is in the active state, the BM-SC initiates the session start procedure.

The MBMS deregistration procedure removes branches of the MBMS delivery tree and any associated MBMS bearer contexts that have been created during the MBMS registration procedure. As depicted in Figure 6.8, the MBMS deregistration procedure comes in two variations, namely a common MBMS deregistration procedure and an BM-SC-initiated deregistration procedure. With the common MBMS deregistration procedure, a downstream node is able to notify an upstream node that it does not need to receive any signalling, session attributes and data for an MBMS bearer service. The deregistration procedure removes the downstream node from the MBMS distribution tree and deletes any relevant MBMS bearer context in the node. The BM-SC-initiated deregistration procedure is usually invoked when a specific MBMS bearer service is terminated, in which case the established MBMS distribution tree is torn down.

There are several causes of the initiation of the common MBMS deregistration procedure. An RNC may initiate the procedure when it notices that it is no longer hosting any UEs for a given MBMS bearer service. Either the SGSN or the GGSN may invoke the procedure when the last MBMS UE context for a particular MBMS bearer service is removed and the list of downstream nodes attribute for an MBMS bearer service is empty or when the last downstream node for an MBMS bearer service deregisters. The detailed steps of the common MBMS deregistration procedure are described in the following:

1. In order to trigger the common MBMS deregistration procedure, the RNC sends an MBMS deregistration request to the SGSN to which it had previously registered for

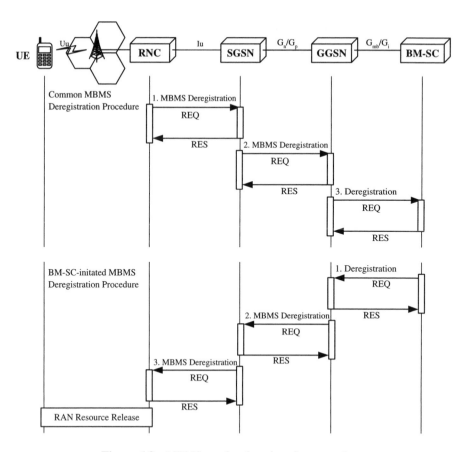

Figure 6.8 MBMS service deregistration procedure

the MBMS bearer service in question. Based on the IP multicast address and the APN provided by the RNC in the request, the SGSN locates the corresponding MBMS bearer context and removes the RNC from its list of downstream nodes. The SGSN confirms the removal by sending an MBMS deregistration response back to the RNC. The RNC can free all relevant MBMS bearer context for that bearer service. Any associated bearer plane resources are released as well.

2. If no UE contexts for an MBMS bearer service exist and the list of downstream nodes attribute of the corresponding MBMS bearer context is empty, the SGSN initiates deregistration by sending an MBMS deregistration request to its upstream GGSN. The GGSN locates the MBMS bearer context with the IP multicast address and the APN provided in the request, removes the SGSN from the list of downstream nodes attribute and confirms the successful deregistration by sending an MBMS deregistration response. The SGSN removes the MBMS bearer context and releases any associated bearer plane resources for the MBMS bearer service.

3. Similarly, if a GGSN determines that the list of downstream nodes for an MBMS bearer service is empty and no MBMS UE contexts for the service exist, it will send a deregistration request to the BM-SC. The BM-SC locates the MBMS bearer context on

the basis of the IP multicast address and the APN provided in the request and removes the GGSN from the list of downstream nodes attribute for the MBMS bearer service. It confirms the removal by sending a deregistration response back to the GGSN. The GGSN deletes the corresponding MBMS bearer context and releases any associated bearer plane resources that may have been established towards the BM-SC.

The MBMS deregistration procedure initiated by the BM-SC removes the complete MBMS delivery tree for a multicast bearer service. It results in the deletion of all MBMS bearer and MBMS UE contexts in all network nodes along the bearer path and releases all associated bearer plane resources. As can be seen in Figure 6.8, the BM-SC triggers the MBMS deregistration procedure for an MBMS bearer service by sending a deregistration request to each GGSN in the list of downstream nodes of the associated MBMS bearer context. The GGSN returns with the deregistration response and deletes all associated MBMS UE contexts. The GGSN also sends an MBMS deregistration message to each node in the list of downstream nodes of the corresponding MBMS bearer context and removes the MBMS bearer context and all associated bearer resources that have been established once it receives the responses from the SGSNs. The SGSNs perform the same procedure with their downstream RNCs. All RAN resources that have been established for the MBMS bearer service are also released.

6.3.3 MBMS Session Control

The MBMS registration procedure only establishes the multicast delivery tree and the necessary MBMS bearer context in the UMTS network. It does not establish the actual bearer resources. In order to allow for fine-grained control of scarce network resources such as radio bearers, MBMS bearer resource requirements are described by different states in the status variable of the MBMBS bearer context. As shown in Figure 6.9, this state model is controlled by the MBMS session start and stop procedures. In stand-by state, no bearer plane resources for the data transfer are required. An active state indicates that bearer plane resources are required.

The MBMS session start procedure is typically initiated by the BM-SC before an MBMS session is scheduled to start. This triggers the establishment of bearer resources along the delivery path. This also includes the set-up of appropriate radio bearers over the air interface, as well as the notification of UEs. As the process may take several seconds

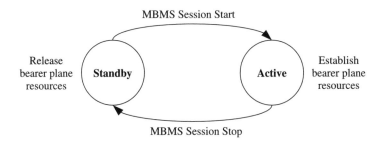

Figure 6.9 MBMS bearer service state model

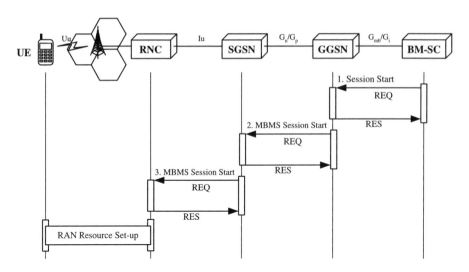

Figure 6.10   MBMS session start procedure

or more, a grace period should be granted prior to sending MBMS session data. This is useful in order to avoid buffering in the network nodes along the bearer path.

Besides setting up bearer resources, the session start procedure is also used for the establishment of a delivery tree and the associated MBMS bearer contexts for MBMS broadcast bearer services. Figure 6.10 illustrates the MBMS session start procedure.

The MBMS session start procedure consists of the following steps:

1. The BM-SC indicates a session start to each GGSN in its list of downstream nodes by sending an MBMS session start request, providing session attributes such as the TMGI, QoS parameters, the MBMS service area, the session duration and so forth in the request. The BM-SC sets the state attribute in the corresponding MBMS bearer context to active. In case of an MBMS bearer service in broadcast mode, the BM-SC also provides a list of downstream nodes that are expected to be part of the delivery tree. The GGSN stores the session attributes in the MBMS bearer context and, in the case of broadcast mode, also adds the list of downstream nodes to the MBMS bearer context. It then sets the state attribute of the MBMS bearer context to active and sends an MBMS session start response back to the GGSN.

2. Similarly, the GGSN distributes the session start indication to each of its downstream SGSNs by sending an MBMS session start request with the previously obtained session attributes. For a broadcast MBMS bearer service, the SGSN creates an MBMS bearer context. The SGSN stores the provided session attributes in the corresponding MBMS bearer context, sets the state attribute to active and provides the TEID of the bearer plane to be used towards the GGSN with an MBMS session start response. The GGSN updates its bearer context entry for the SGSN accordingly.

3. The SGSN distributes the MBMS session start request to each of the RNCs that have an entry in the list of downstream nodes attribute, providing all relevant session attributes. In case of an MBMS broadcast bearer service, the RNC first creates a corresponding MBMS bearer context. The RNC then adds the session attributes to the MBMS bearer

context, sets the state attribute to active and provides the TEID for the bearer plane to be used to the SGSN in the MBMB session start response. The SGSN updates its bearer context entry for that RNC accordingly. As a result, the RNC establishes the radio bearer resources according to the MBMS data transfer parameters stored as session attributes.

Figure 6.11 illustrates the MBMS session stop procedure that is typically triggered to release the bearer service resources after an MBMS session is terminated. The MBMS session stop procedure can also be invoked during an MBMS session if a period of data inactivity is sufficiently long to justify a release of the bearer service resources. The procedure is detailed in the following:

1. The BM-SC sends an MBMS session stop request to all downstream GGSNs in the delivery tree to indicate that a session has terminated and bearer resources can be released. It also sets the state attribute of the MBMS bearer context to standby. The GGSN replies with an MBMS session stop response and sets its own state attribute to standby. The BM-SC may also trigger the generation of charging data for MBMS multicast bearer services.
2. The GGSN sends an MBMS session stop message to each SGSN in the list of downstream nodes of the MBMS bearer context and sets the respective state attribute to standby to initiate the release of bearer resources towards the SGSNs. In the case of a MBMS broadcast bearer service, the GGSN deletes the MBMS bearer context.
3. The SGSN replies with an MBMS session stop response and releases the TEID and bearer plane resources on which it was receiving MBMS session data from the GGSN. It sends an MBMS session stop request to each downstream RNC it was serving and sets the state attribute of the MBMS bearer context to standby in order to initiate a release of bearer resources towards the RNCs. If the MBMS bearer service is in broadcast mode, the SGSN deletes the respective MBMS bearer context.

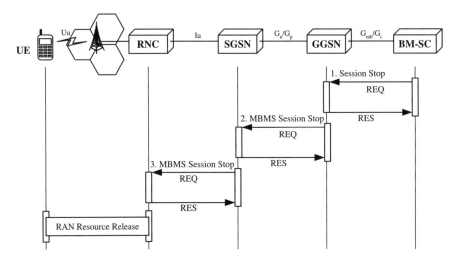

Figure 6.11   MBMS session stop procedure

4. The RNC replies with an MBMS session stop response and initiates the release of radio bearers and the bearer resources towards the SGSN. In the case of an MBMS broadcast bearer service, the MBMS bearer context is deleted.

Besides starting and stopping sessions, MBMS also supports the update of sessions that are in progress. Updating ongoing sessions is handled by the MBMS session update procedure. The MBMS session update procedure has two purposes. When initiated by the SGSN during an ongoing MBMS multicast session, the SGSN informs the RNC that UEs are moving to a new routing area or that a routing area is no longer hosting interested UEs. The other purpose of the MBMS session update procedure is to propagate a service area change within the delivery tree of an MBMS broadcast bearer service. In this case, the BM-SC initiates the procedure as depicted in Figure 6.12 in order to update the MBMS service area attribute and potentially also the list of downstream nodes of the corresponding MBMS bearer contexts. The procedure is described briefly in the following:

1. The BM-SC sends a session update request to all downstream GGSNs of the MBMS broadcast bearer service. The request identifies the affected MBMS broadcast bearer service and provides the new configuration of the new service area in terms of a new *list of downstream nodes* and *MBMS service area* attributes. The GGSN stores the new attributes in the corresponding bearer context and sends a session update response back to the BM-SC.
2. The GGSN compares the new list of downstream nodes attribute with the former one. It sends an MBMS session start request to each newly added SGSN and an MBMS session stop request to each SGSN that has been removed, and forwards the MBMS session update request to the SGSNs that were already in the list.
3. An SGSN receiving the MBMS session update request updates the session attributes of the MBMS bearer context accordingly and verifies whether they have changed.

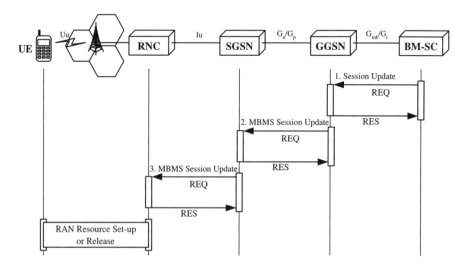

Figure 6.12   MBMS session update procedure

It implements any changes by sending MBMS session management messages to its downlink RNCs, equivalently to the actions taken by the GGSN in step 2. As a result of the session update request, radio bearer resources may be established or released in the RAN.

6.3.4 MBMS Service Request

Over the air interface, MBMS users can be served either via dedicated PTP or shared PTM radio bearers. The most appropriate choice between both options depends on the number of MBMS users in a given cell. The UE linking procedure informs UTRAN only of MBMS receivers that are in the PMM-CONNECTED state. Before and during a session, only a subset of these receivers may be known to UTRAN. UTRAN may initiate the MBMS service request procedure in order to know exactly how many MBMS receivers it is serving. The MBMS service request procedure is illustrated in Figure 6.13 and essentially triggers a percentage of receivers to change from the PMM-IDLE to the PMM-CONNECTED state. Details of the procedure are provided in the following:

1. A PMM-IDLE UE learns from the UTRAN over the MCCH that an MBMS service request needs to be performed. The UE determines whether it must perform an RRC connection establishment in order to send a service request to the SGSN.
2. Assuming that the UE must perform the MBMS service request procedure, the UE sends an MBMS service request message to the SGSN, indicating either MBMS multicast or MBMS broadcast reception as the service type.

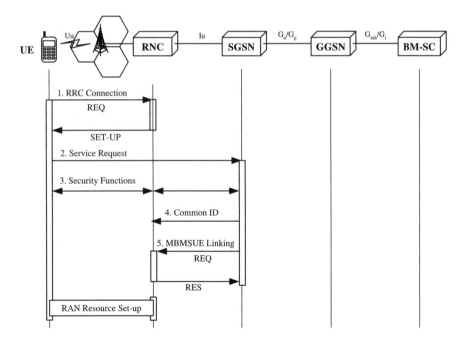

Figure 6.13 MBMS service request procedure

3. The SGSN may perform security functions to authenticate the UE.
4. The SGSN provides the RAN with the IMSI of the UE.
5. By means of the UE linking procedure, the SGSN provides the RAN with information of all MBMS services that the UE has currently activated.

The SGSN then uses the provided information to *count* the number of receivers. Based on a given threshold, the SGSN may choose to establish or modify the radio bearers accordingly.

6.4   Routing on the MBMS Bearer Path

In today's UMTS networks, forwarding of user data from the GGSN up to the RNC is realized by IP routers in the backbone of the operator. User data is usually tunnelled from network node to network node using GTP. More specifically, the transport of user data along the bearer path is realized by the GTP-U protocol, while the necessary control plane signalling between the GSNs is realized by the GTP-C protocol. Control plane signalling between the SGSN and the RNC is carried out via RANAP. GTP-U supports the transport of data packets from a variety of different user plane protocols such as IPv4, IPv6 or PPP.

Two options exist in order to support the transport of MBMS user service data along the MBMS bearer path (3GPP, 2003). With the first option, MBMS user service data is tunnelled by means of IP unicast along the bearer path of an MBMS bearer service from network node to network node. With the second option, IP multicast is used within the operator's backbone to tunnel MBMS user service data along the bearer path.

Figure 6.14 illustrates the first option, which relies on unicast tunnelling. Network nodes send data packets directly addressed to each other, and IP routers in the operator's backbone route packets on the basis of the IP unicast address of a particular destination node. A Content Provider (CP) in an external IP-based packet data network sends user service data addressed to the IP multicast address X. It is assumed that an MBMS multicast bearer for this user service has already been established, and that the associated MBMS bearer contexts exist in the network nodes along the bearer path. The dotted lines between the network nodes indicate the logical delivery tree for the MBMS bearer service. For simplicity, the BM-SC, usually located between the external network and the GGSN, is omitted from Figure 6.14.

The GGSN is able to look up the corresponding MBMS bearer context on the basis of the IP multicast address of the incoming user data packet. For each entry in the list of downstream nodes, the GGSN retrieves the corresponding TEID and the unicast IP address of the SGSN. It duplicates the user data packet for each SGSN and encapsulates the user data packet into a GTP packet with the retrieved TEID and then sends the packet to the respective SGSNs. IP routers along the path forward the packets to the destination SGSNs. An SGSN receiving an incoming MBMS user data packet via the GTP-U tunnel first retrieves the corresponding MBMS bearer context and then looks up the corresponding IP addresses and TEID of the RNCs in the list of downstream nodes of the MBMS bearer context. A copy of the packet is then sent via GTP tunnels to each of the RNCs. The routers along the path forward the data on the basis of the IP unicast

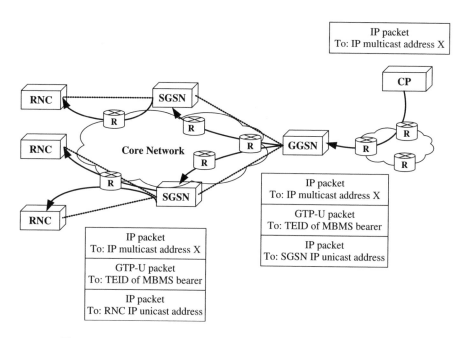

Figure 6.14   Unicast routing of user service data in the core network

addresses of the RNC. The RNCs then further distribute the user data packet over the corresponding radio bearers. This approach is much more resource efficient than traditional UMTS unicast delivery, as not one copy per user but only one copy per downstream node is forwarded by each network node. This approach is not optimal, as network nodes still need to send one copy per downstream node. Also, routers may forward redundant data on the common paths towards the downstream nodes.

Figure 6.15 illustrates the approach in which IP multicasting is used to distribute the data to the network nodes along the bearer path. With this approach, each GGSN and SGSN uses a separate IP multicast address within the operator's backbone to forward data to its downstream nodes. Downstream nodes are informed of this multicast address during the establishment of the MBMS bearer plane and join the corresponding multicast group at their local multicast-enabled router. Instead of tunnelling the received MBMS user data packet to each SGSN in the list of downstream nodes separately, the GGSN only sends a single copy, addressed to the localized multicast group to which its downstream nodes are subscribed. Multicast routers along the path replicate the data packet as required to the respective SGSNs. The SGSN looks up the respective MBMS bearer context on the basis of the TEID and tunnels only a single copy addressed to the IP multicast group to which its downstream RNCs are subscribed. Compared with the first approach, only a single copy of a user data packet needs to be sent by a network node. IP routers in the network do not need to transmit redundant copies of the same user data packet on common paths. This approach comes at the cost of additional complexity in maintaining several internal IP multicast groups for a single MBMS bearer service.

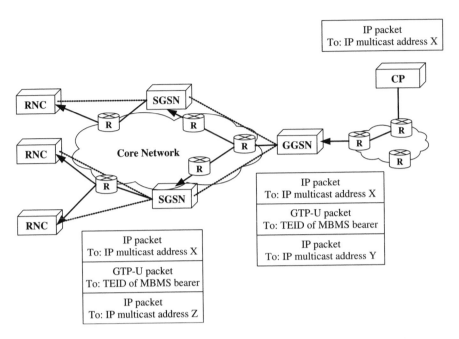

Figure 6.15   Multicast routing of user service data in the core network

6.5   MBMS User Services

Besides MBMS bearer services, MBMS provides a service layer that is required for the delivery of MBMS-based services in the network. MBMS user service entities are located at the top of this service layer. These entities serve as the interface of an application or service towards the end-user. MBMS user services can be offered with informative material describing the content as well as other characteristics and scheduling information of the service. This information gives the end-user a basis for deciding whether an offered service is of interest. User service discovery and announcement mechanisms provide this information to the user. MBMS user services rely on security functions provided as part of the MBMS service layer in order to ensure that only authorized users are able to access the content of the MBMS user services. Authentication and content encryption are examples of such security functions.

An MBMS user service entity can consist of multiple distinct multimedia objects or streams. Each of the multimedia objects or streams may impose different requirements on the transport of content. For example, live streaming of a Premiership football match has different requirements on the underlying transport compared with the delivery of an electronic newspaper or magazine. The MBMS service layer offers two different delivery methods, namely download delivery and streaming delivery. These delivery methods provide a set of protocols and procedures, such as FEC techniques, file repair or delivery verification, that are tailored to the necessary delivery characteristics of the user service data.

An MBMS download or streaming session in turn makes use of one or more MBMS or unicast bearer services provided by the MBMS network layer. Figure 6.16 shows several

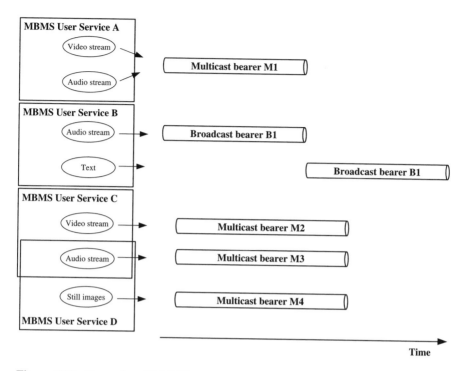

Figure 6.16   Examples of MBMS user service sessions using different MBMS bearers

possibilities for delivering streaming multimedia components of an MBMS user service. An audio and a video stream component can be transported as part of a single MBMS streaming session that makes use of an MBMS multicast or an MBMS broadcast bearer. Audio and video components can also be transported via separate MBMS multicast bearers in the same MBMS streaming session or in successive sessions. Another user service may make use of the same MBMS bearer service in order to combine the audio stream with still images, with the still images being transported in an MBMS download session with a separate MBMS multicast bearer. In general, it is recommended to transport MBMS download or streaming sessions belonging to the same MBMS user service with the same MBMS bearer service. In Release 6, MBMS broadcast and multicast bearers cannot be combined in a single MBMS session.

   The following sections introduce the different protocols and mechanisms provided by the MBMS service layer. Firstly, the MBMS streaming and download delivery methods are presented. Then, the most important user service support functions are described.

6.5.1   MBMS Streaming Delivery Method

The MBMS streaming delivery method provides support for the delivery of continuous multimedia data such as speech or audio and video streams. The delivery method is useful for MBMS user services that require the delivery of scheduled multicast and broadcast streaming content. MBMS multicast and broadcast bearer services as well as unicast UMTS bearer services can be used to deliver these services.

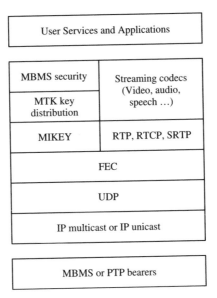

Figure 6.17 Protocol stack for streaming delivery method

Figure 6.17 illustrates the protocol stack of the MBMS streaming delivery method. The main transport protocol used is RTP (Group *et al.*, 1996). RTP operates on top of UDP and is commonly used for the real-time transfer of audio and video streaming data in data networks. RTP provides support for packet sequencing and time-stamping such that packet ordering, synchronization and jitter calculations can be performed at the receiver. RTP is used in conjunction with the Real-Time Transport Control Protocol (RTCP). RTCP is used to provide feedback on the QoS of the RTP session by means of out-of-band control information. As MBMS bearer services are unidirectional, only downlink messages of RTCP are permitted. RTCP reports that would usually be sent back to the source are suppressed at the receiver. This is achieved by configuring RTCP accordingly during service discovery. The multimedia codecs and the RTP payload formats are compliant with the formats specified by the UMTS Packet-Switched Streaming (PSS) service (3GPP, 2007i).

The Secure Real-Time Transport Protocol (SRTP) profile provides support for authentication, data integrity and encryption if secure transmission of session data is required. The distribution and management of encryption keys is handled by the Multimedia Internet Keying (MIKEY) protocol, which is used in conjunction with SRTP. More details on the security functions of MBMS are presented in Section 6.5.6. The MBMS streaming delivery method offers the use of an FEC mechanism as an option in order to increase the reliability of the service. With FEC, UDP packets containing either RTP, RTCP, SRTP or MIKEY packets as payload are encoded at the BM-SC into FEC source blocks. Also, the desired amount of recovery symbols is generated. The protected source data and recovery symbols are packed separately into FEC source and FEC recovery packets respectively. The FEC source and recovery packets are then sent over the MBMS bearer service. At the receiver, the FEC mechanism recovers the original packets directly from the FEC source packets and buffers them for a minimum amount of time to allow time for FEC repair.

The FEC decoder tries to reconstruct the initial FEC source block using the FEC source and repair packets. If insufficient FEC source and repair packets are received, only the original packets that were received as FEC source packets will be available. The rest of the original packets are lost. The exact operation of the FEC mechanism is described in (3GPP, 2007e).

An MBMS streaming method optionally provides support for the collection and exchange of an MBMS Quality of Experience (QoE) metrics during an MBMS streaming session. The QoE metrics are captured at the MBMS client application and reported to the MBMS streaming server that provides the MBMS session content. Examples of QoE metrics are corruption duration, initial buffering and rebuffering duration, successive loss of RTP packets or frame and jitter duration. The feature has to be explicitly enabled by the MBMS user service by providing a corresponding entry in the service description during service announcement. An MBMS client application that supports this feature performs the quality measurements in accordance with the measurement definitions, aggregates them into QoE metrics and reports them back to the MBMS streaming server via the content reception reporting procedure. The content reception reporting procedure is described in Section 6.5.5.

6.5.2 MBMS Download Delivery Method

The MBMS download delivery method provides support for reliable delivery of data files. The delivery method is suitable for any MBMS user service that needs to distribute MBMS content in the form of data files, binary data, still images or metadata, which is used for service announcements. The MBMS download delivery method allows for the transfer of an arbitrary number of files within a single data transfer session.

Figure 6.18 illustrates the protocol stack of the MBMS download delivery method using a PTM MBMS bearer service. The main transport protocol is the FLUTE protocol (Paila *et al.*, 2004). FLUTE provides reliable PTM data transfer over unidirectional links and makes use of an FEC building block to increase the reliability of data delivery. FLUTE cannot provide absolute reliability. In MBMS, the file repair procedure is used to guarantee reliability. Further details of this procedure are given in Section 6.5.4. In FLUTE, FDTs are used to identify different files within a FLUTE transmission. The download delivery method typically associates one FDT instance with a single MBMS download delivery session. An MBMS client application on the UE has three ways to receive files from an MBMS download delivery session. In *promiscuous* mode, the MBMS client application instructs the local FLUTE entity to receive all available files during a session. In *one-copy* mode, the local FLUTE entity is instructed of one copy of one or more specific files, identified by a file URI. The FLUTE entity does not have to be connected to the entire MBMS download session, but can exit as soon as all required files are received. The file URIs are typically known from session descriptions provided by the MBMS session announcement. In *keep updated* mode, the FLUTE entity is instructed to receive one or more files as well as updates of these files. Details on the use of FLUTE in MBMS can be found in (3GPP, 2007e).

The protocol stack utilized for reliable download delivery over PTP bearer services is depicted in Figure 6.19. MBMS user service data may be transported reliably via unicast UMTS bearer services when MBMS bearers are not available. In such a case, the delivery is based on the procedures defined by the Open Mobile Alliance (OMA)

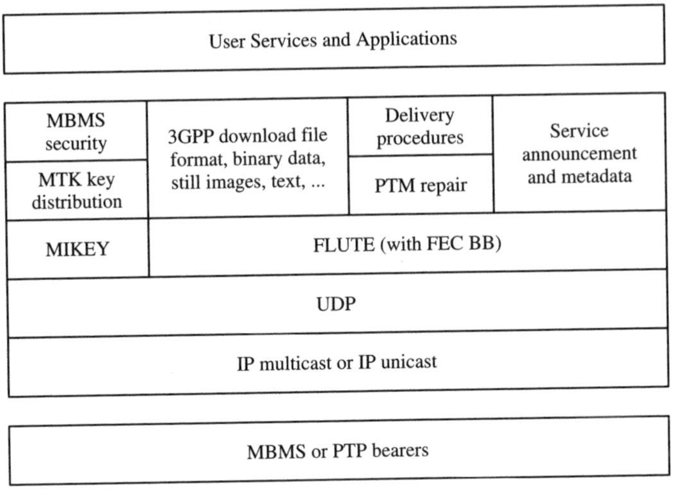

User Services and Applications			

MBMS security	3GPP download file format, binary data, still images, text, ...	Delivery procedures	Service announcement and metadata
MTK key distribution		PTM repair	
MIKEY	FLUTE (with FEC BB)		
UDP			
IP multicast or IP unicast			

MBMS or PTP bearers

Figure 6.18   Protocol stack for PTM reliable download delivery

User Services and Applications

MBMS security		Delivery procedures		Service announcement and metadata
MSK key distribution	Registration	Reception reporting	PTP repair	
MIKEY	HTTP			
UDP	TCP			
IP unicast				

PTP bearers

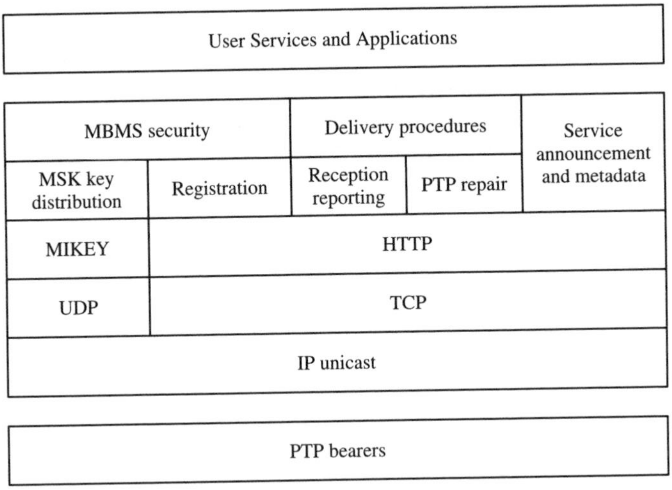

Figure 6.19   Protocol stack for PTP reliable download delivery

push specification (OMA, 2007). The UE has to register its MSISDN with the BM-SC to receive the download session using OMA push. Registration and deregistration is handled by the HTTP protocol. The protocol stack in Figure 6.19 is also used for PTP file repair by means of the file repair procedure described in Section 6.5.4, for interactions with security functions or for the dissemination of encryption keys.

6.5.3   MBMS User Service Announcement and Discovery

MBMS user service announcements provide MBMS service providers with the means to advertise the list of available user services and service bundles along with information on these. The information contained in the service announcements not only allows users to select services but also allows the selected services to be established.

User service descriptions are provided as metadata fragments. These metadata fragments, which are uniquely identifiable blocks of metadata, provide details of the MBMS user services. The metadata fragments typically contain descriptive information on the MBMS user services or user service bundles, session information, information required for the associated delivery methods, information on service protection and details of the FEC data stream, if necessary. The metadata fragments are accompanied with metadata envelope objects, providing metadata management information to a particular metadata fragment. If transmitted over a download bearer service, both should be transported as file objects in the same download session, either as cross-referenced files or embedded in a single file object. Further details on the format of the metadata fragments can be found in (3GPP, 2007e).

6.5.4 File Repair Procedure

The file repair procedure is used to recover lost or corrupted file fragments. With this procedure, each receiver of an MBMS download delivery session identifies the missing data fragments and requests these from a file repair server after the end of the session. The file repair server then provides the missing data either with unicast or MBMS bearer services.

The potentially large number of receivers that may wish to recover missing or corrupted fragments of a file can easily lead to scalability issues. Feedback implosion leading to congestion on the up- and downlink as a result of serving many simultaneous requests with the repair traffic are some of the common issues that may arise. Furthermore, the file repair server may be overwhelmed with a large number of file repair requests. In order to avoid such problems from occurring, file repair ideally should be spread over an extended period of time and serviced by file repair servers.

The file repair request procedure requires each MBMS client to calculate a random back-off time prior to sending the repair request. The random back-off time is the period of time the MBMS client should wait until sending the request. The back-off time interval can be adapted according to the expected receiver population. An MBMS client then selects a file repair server randomly from a list of servers and sends the repair request after the back-off time has elapsed. The repair request is sent with unicast UMTS bearers. The file repair server receiving the request sends a file repair response message. Several options for providing the repair data exist. The file repair server may decide to provide the requested repair data with a repair response message, may redirect the client to another MBMS download session or may redirect the request to another repair server in order to balance the repair load.

6.5.5 Reception Reporting Procedure

The reception reporting procedure can be initiated by an UE in order to confirm the successful reception of service content. The procedure is essential for realizing service charging based on the successful reception of content.

For the MBMS download delivery method, reception reporting is used to report on the successful reception of one or more data files received during an MBMS session.

Each UE participating in the session may have to report if the reception confirmation of data files is required for the MBMS user service. With the streaming delivery method, reception reporting is used to report on the statistics of the streaming session. In this case, the BM-SC may specify a percentage of receivers that are required to report on the statistics of the streaming session. This ensures that the resulting reporting traffic is minimized.

A UE must determine whether it needs to initiate the reporting procedure when an MBMS session has terminated. The need for reception reporting is indicated as part of the delivery procedure description provided with the service description of the MBMS user service. If only a subset of receivers is required to report, the UE must determine whether it is part of the subset. If so, the UE first computes the reporting time randomly based on the back-off time interval and then sends the reception report to the reception report server via a unicast bearer service. The reception report service acknowledges the reception report with a reception report response message.

6.5.6 MBMS Security

The security functions for MBMS enable the secure delivery of MBMS service data to a set of eligible users. For this purpose, mechanisms for the authentication of MBMS users, for the distribution of security keys and for the protection of MBMS user service data have been introduced as part of the MBMS service layer. Not all MBMS user services require the protection of transmitted data.

An MBMS user service is provided in the form of one or more MBMS streaming or MBMS download sessions. Streaming and download sessions may be transported over one or more MBMS bearer services or unicast bearer services. A streaming session is composed of one or more RTP sessions. Likewise, an MBMS download session is composed of one or more FLUTE channels. MBMS security is applied in order to protect the RTP sessions and the FLUTE channels. As such, the security mechanisms remain independent of the bearer service selected for data transport.

The transfer of both RTP sessions and FLUTE channels are typically protected with symmetric keys, referred to as MBMS Traffic Keys (MTKs). MTKs are shared by the BM-SC and the UEs. The protection of data applies on an end-to-end between the BM-SC and UE and is performed by the session and transmission function in the BM-SC. Depending on the MBMS user service, different protection methods that ensure either confidentiality or integrity or both may be required. Key identification information is included with the protected data in order to allow the UE to determine the MTK used for the protection. This is essential for decryption. The delivery of MTKs to the relevant MBMS users is protected by Multicast Service Keys (MSKs). The delivery of MSKs is in turn secured by the MBMS User Key (MUK). Finally, the MBMS Request Key (MRK) is used to authenticate an UE when performing MBMS user service signalling. Both the MRK and the MUK are shared keys that are derived from keys obtained via a Bootstrapping Service Function (BSF) of the generic bootstrapping architecture (3GPP, 2007c).

A high-level view of the MBMS security architecture is shown in Figure 6.20. Most of the security functions reside within the BM-SC and the UE. The BM-SC provides key management in the form of key request and key distribution functions. The key request function is responsible for deriving the MRK and the MUK for a subscriber, based on a

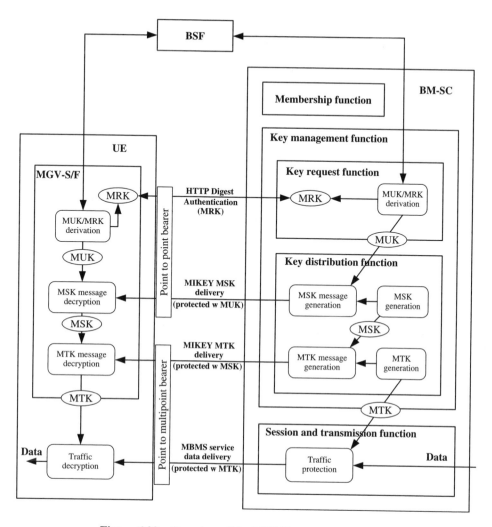

Figure 6.20   Overview of the MBMS security architecture

key obtained by the BSF. It performs the authentication of an UE via HTTP digest and subsequently handles the MBMS user service registration, MBMS user service deregistration and the MSK request procedures from the UE. It provides the key distribution function with the MUK of a user and performs subscription checks from the membership function in the BM-SC. The key distribution function generates both the MSK and the MTK and is responsible for distributing these keys to the UE. The session and transmission function uses the MTK provided by the key distribution functions and performs the protection of MBMS streaming and download session data. The membership function verifies whether a particular user is authorized to register for MBMS user service, receive keys for a session or establish an MBMS bearer service. On the UE side, the MBMS key generation and validation functions maintain the MBMS keys in a secure storage space referred to as the MBMS Key Generation and Validation Storage (MGV-S). Analogue to the BM-SC, the MRK and MUK are generated from a key obtained by the BSF. The UE is able to register for MBMS user service sessions and request the respective MSKs and MUKs. The MSK is delivered via the MIKEY protocol over a PTP bearer service and decrypted with the MUK. The MTKs are typically provided via the MIKEY protocol over an MBMS bearer service and decrypted with the help of the MSK. The MTK is then used for the decryption of MBMS user service data.

6.6 Summary

This chapter has described the protocols and mechanisms used to deliver multicast and broadcast services in UMTS networks. The MBMS service architecture can be divided into the MBMS service and network layers. MBMS user services make use of the different delivery methods provided by the MBMS service layer. The service layer provides additional support mechanisms such as service announcement and security functions. MBMS user services rely on different bearer service capabilities provided by the MBMS network layer in order to transport MBMS user service data and service-related information. Only two of the four UMTS QoS classes are supported for MBMS bearer services. MBMS bearer services of the streaming class provide support for a guaranteed bit rate and are most suited for the transport of MBMS user services that provide some form of real-time content delivery. MBMS bearer services of the background class are most suited for the delivery of data files. The MBMS distribution tree has to be constructed in such a way that it can provide the same QoS on all of its branches. Network nodes along the bearer path maintain two types of state information, namely MBMS bearer contexts and MBMS UE contexts. MBMS bearer contexts contain the state information required for forwarding service data, whereas MBMS UE contexts contain user-specific information required for MBMS in multicast mode. The management of MBMS bearer services is governed by a set of mechanisms referred to as MBMS procedures. These MBMS procedures specify how and when MBMS bearer services and the associated bearer state are created or released within the UMTS network and how bearer service paths and the associated network resources are managed during service provisioning. The MBMS service layer provides two delivery methods, namely streaming delivery and download delivery. The streaming delivery method provides support for the delivery of continuous multimedia data such as speech or audio and video streams. The MBMS download delivery method provides support for the reliable delivery of data files. Several support mechanisms in the service layer were discussed. The MBMS user service announcement and discovery mechanisms allow providers to advertise MBMS services and upcoming sessions. The file repair procedure provides guaranteed reliability by allowing receivers to recover from lost or corrupted data during an MBMS download session. The reception reporting procedure can be used in order to report the successful download of a file or provide statistics during an MBMS streaming session. The MBMS security architecture finally provides the means for secure delivery of MBMS service data.

7

Protocols and Mechanisms for BCMCS

7.1 Introduction

This chapter extends the initial overview of BCMCS provided in Chapter 5 and presents the protocols and mechanisms of BCMCS in detail. Apart from providing broadcast and multicast capabilities in CDMA2000 networks, BCMCS also introduces a service framework consisting of different support functions. This service framework is depicted in Figure 7.1. Similarly to the service framework in MBMS, the functions provided by the BCMCS service framework can be roughly grouped into two layers, a service layer and a network layer. The service layer comprises the functions necessary for the set-up and management of BMCMS broadcast sessions, while the BCMCS network layer provides the BCMCS bearer services for efficient delivery of BCMCS flows as well as several functions for the management of bearer services.

Architecturally, the service-layer functions are realized by the BCMCS controller, BCMCS content server and BCMCS subscription manager on the network side, and by the respective application entities at the MS or AT on the user side. Within the BCMCS service architecture, multicast and broadcast content is provided as BCMCS broadcast programmes, which are equivalent to user services in MBMS terminology. The service flows of a broadcast programme are provided via the BCMCS content server, which serves as the network entry point for BCMCS content. A content provider is able to control a BCMCS service session via the BCMCS controller. Management of user subscriptions is handled by the BCMCS subscriber profile manager. The service layer also provides mobile users with the capability to discover BCMCS broadcast programmes and is responsible for securing transmissions, so that only authorized subscribers are able to receive the given content. It should be noted that the current BCMCS service layer only provides support for streaming delivery, unlike the MBMS service layer which also supports reliable file download.

Multicast in Third-Generation Mobile Networks   Robert Rümmler, Alexander Gluhak and A. Hamid Aghvami
© 2009 John Wiley & Sons, Ltd

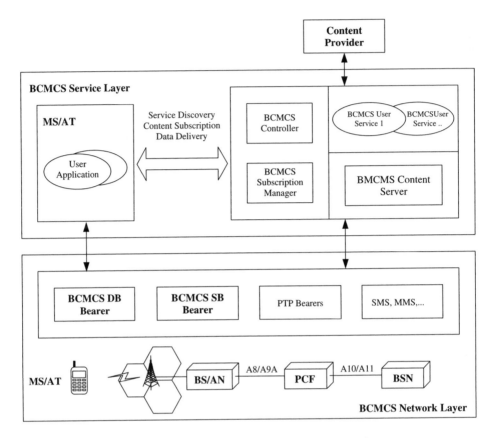

Figure 7.1   An overview of the BCMCS service architecture

The BCMCS network layer facilitates the delivery of multicast and broadcast content and the interactions required for service discovery and content subscription. This is done by providing the underlying transport in the form of different bearer services. These bearer services include the BCMCS Dynamic Broadcast (DB) and Static Broadcast (SB) bearer services, the conventional unicast bearer services or other alternative forms of message delivery such as SMS or MMS.

The rest of the chapter is organized as follows. Firstly, the BCMCS broadcast bearer architecture is introduced, followed by a detailed description of how the BCMCS bearer plane is managed. Then, an overview of the protocols and mechanisms of the BCMCS service layer is provided. The chapter finally closes with a summary of the presented content.

7.2   BCMCS Bearer Path Architecture

The BSN is a new functional node introduced as part of the BCMCS standard. The BSN can be seen as a logical extension of the PDSN, handling the addition or removal of IP multicast flows towards the RAN. Figure 7.2 illustrates the bearer path architecture for BCMCS bearer service.

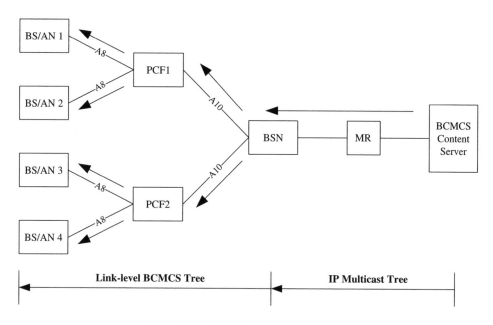

Figure 7.2   BCMCS bearer architecture

The interfaces for the transport of data and control signalling between the logical network nodes remain largely unchanged. The A8 and A10 interfaces depicted in Figure 7.2
are used for the transport of BCMCS flows, while the A9 and A11 interfaces are used
for the exchange of control signalling. The protocol stacks across all interfaces remain
unchanged apart from some modifications required for the delivery of BCMCS flows. In
order to transport BCMCS flows, certain restrictions on the use of the Generic Routing
Encapsulation (GRE) protocol have been introduced over the A8 and A10 interfaces, with
several new procedures supporting these restrictions. Likewise, several procedures have
been added for protocols over the A9 and A11 interfaces to support the management of
the BCMCS bearer plane. These signalling procedures are described in the next section
of this chapter. More details on these interfaces and the respective BCMCS enhancements
can be found in (3GPP2, 2007b) and (3GPP2, 2006b).

As can be seen from Figure 7.2, the BCMCS bearer architecture is essentially a two-
tiered multicast architecture. The first tier relies on network-level IP multicast, in which
BCMCS flows are disseminated by IP multicast routers from the BCMCS content server
to BSN(s) via IP multicast delivery trees. Different BCMCS flows are typically sent
out to different IP multicast groups. In the second tier, BCMCS flows are distributed
by broadcast bearer connections over the A10 and A8 interfaces, essentially forming a
link-level multicast tree. Both broadcast bearer connections are unidirectional, carrying
data over the A10 interface from a BSN to the PCF and over the A8 interface from the
PCF to the BS or AN. A PCF that is part of the delivery tree of a particular BCMCS flow
receives a flow on at most one A10 broadcast connection. Likewise, different BCMCS
flows have to be carried on separate A10 broadcast connections to a PCF and cannot
be combined on a single A10 broadcast connection. The same rules apply to carrying
BCMCS flows to the BS or AN via the A8 broadcast connection.

7.3 BCMCS Bearer Service Management

For dynamic broadcast bearer services, the bearer paths are managed by bearer service procedures that are part of the control plane of the bearer service. Bearer service procedures operate in a distributed fashion over the different logical network nodes along the bearer path and are responsible for the establishment, maintenance and release of the associated data plane.

With the BCMCS registration procedure, an MS or AT notifies the RAN that it intends to receive one or more IP multicast flows identified by a given set of BCMCS flow IDs. This procedure may trigger the RAN session discovery, which is used by the RAN to obtain session-related information about a requested flow so that the session path can be established. The BSN session information update procedure is used by the BSN to inform downstream nodes of updated BCMCS flow session information. Setting up and tearing down a bearer path for a BCMCS flow is managed by the BCMCS bearer establishment and BCMCS bearer release procedures.

7.3.1 BCMCS Registration and RAN Session Discovery

The BCMCS registration procedure is used by an MS or AT to inform the network about its interest in receiving one or more BCMCS flows. The BCMCS registration procedure is very similar to the MBMS multicast service activation procedure. The BCMCS registration procedure differs from the procedure for multicast service activation in MBMS in two ways. Firstly, it does not make use of IGMP or MLD signalling for group management; instead, BCMCS-specific signalling is used. Secondly, signalling is used to inform the RAN and not the core network, as is the case with MBMS. If an MS or AT is the first one to register for a BCMCS flow at the BS or AN, no session-related information for the BCMCS flow will be available at the BS or AN. As a result, the BS or AN then triggers the RAN session discovery procedure to obtain the relevant BCMCS session information from its upstream nodes.

Figure 7.3 illustrates the signalling flows for the BCMCS registration procedure. The signalling messages over the air interface are described informally, without adhering to the naming and parameterization defined in the standard. Furthermore, the interactions between the BCMCS Controller (BCMCS-C), which are based on the Remote Authentication Dial-In User Service (RADIUS) or Diameter protocol, are somewhat simplified in that the involvement of the serving AAA is not shown. These simplifications apply to all figures in this chapter. The signalling steps shown in Figure 7.3 are described in detail as follows:

1. An MS or AT has been informed of the BCMCS flows through service discovery and receives information on service availability as well as relevant radio parameters by means of network overhead messages.
2. The MS or AT initiates BMCMS registration by sending a BCMCS registration message to its BS or AN, transmitting the identifiers of the BCMCS flows and optionally providing a corresponding authentication signature.
3. With 1xRTT, the BS performs a location update procedure with the MSC. With high-rate packet data systems, for which session control and mobility management

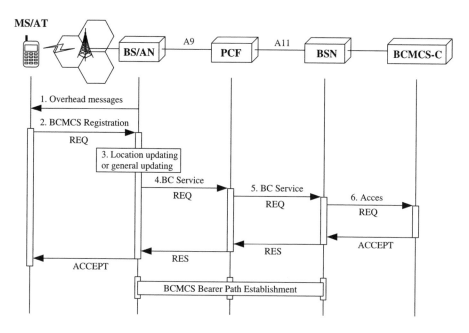

Figure 7.3 BCMCS registration procedure

is not performed in the AN, the AN performs a general update procedure with the PCF.

4. If the BS or AN does not hold session information for the BCMCS flow, it initiates the RAN session discovery procedure by sending a BC service request message to the PCF in order to request the BAK authorization and/or session information for the BCMCS flow. If the PCF holds the request information and can perform the BAK authorization, it immediately returns a BC service response message to the requesting BS or AN.

5. If it does not hold RAN session information, the PCF requests the information as well as user authorization for the BCMCS flow from an upstream BSN by sending the BC service request message. The BSN replies to the request if it holds this information with a BC service response.

6. A BSN that does not yet hold this information for the requested BCMCS flow or requires user authorization sends a RADIUS access request to the BCMCS-C. The BCMCS-C performs user authorization if required and provides the RAN session information for the BCMCS flow in the RADIUS access accept message to the requesting BSN.

7.3.2 BCMCS Session Information Update

The BCMCS session information update procedure is used to provide the RAN with an update of RAN session information for a BCMCS flow. The procedure is typically initiated by the BCMCS-C and can initiate the establishment of a BCMCS bearer path. Figure 7.4

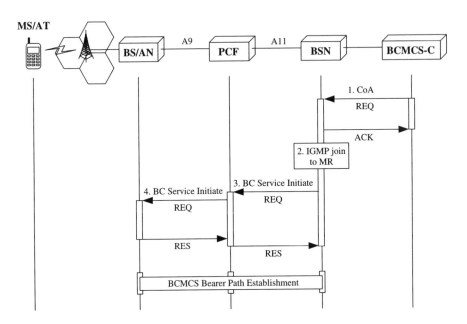

Figure 7.4 BCMCS session information update procedure

illustrates the signalling flow of the procedure. A full description of the signalling steps is given below:

1. The BCMCS-C sends a *CoA request message* to the BSN, typically providing the full set of BSN and RAN session information for a BCMCS flow. The BSN processes this message and replies with a *CoA acknowledge message*.
2. If the information for a BCMCS flow is provided, the BSN joins the IP multicast group via IGMP or MLD signalling with its multicast router.
3. The BSN may send a *BC service initiate request* message to one or more PCFs over the A11 interface, providing the received BCMCS RAN session information.
4. The PCF further distributes the information to its downstream BSs or ANs with a *BC service initiate request* over the A9 interface. The respective BSs or ANs respond with a *BC service initiate response*. The PCFs return a *BC service initiate response* to the initiating BSN.

The BCMCS session information update procedure may result in the establishment of a bearer path for the BCMCS flow. The establishment of a bearer path is described in the next section.

7.3.3 BCMCS Bearer Set-Up

The set-up of a BCMCS bearer path is typically initiated by the BS or AN using the BCMCS bearer establishment procedure. This procedure establishes the link-level multicast tree of the BCMCS bearer path. Figure 7.5 shows the signalling flow involved.

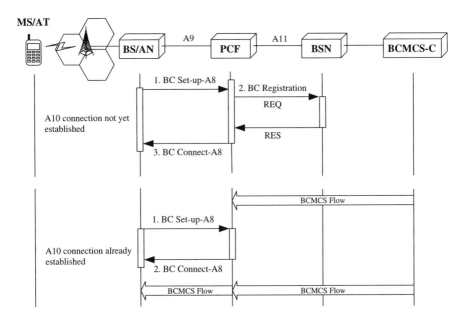

Figure 7.5   BCMCS bearer set-up procedure

The upper half of the figure shows the case where no previous A10 data plane exists towards the PCFs. The lower part of the figure illustrates the case where the PCF is already serving a different BS or AN with the same BCMCS flow. The signalling flow in Figure 7.5 consists of the following steps:

1. The BS or AN may have the relevant RAN session information for a BCMCS flow and determines (for example, by means of a BCMCS user registration) that the transmission of the BCMCS bearer flow is required. It initiates the establishment of the bearer plane for the BCMCS flow by sending a *BC set-up-A8* message via its A9 interface to its PCF. If the PCF is already providing the BCMCS flow to another BS or AT (which means that it already has a corresponding A10 connection established), it immediately returns a *BC connect-A8* message to establish the requested A8 connection and start forwarding the BCMCS flow towards the BS or AN.
2. If no A10 connection exists for the BCMCS flow, the PCF sends a *BC registration request* message with a lifetime attribute set according to the expected session duration to the BSN. This signalling message requests the establishment of the A10 connection. The BSN returns a *BC registration response* to complete the establishment of the A10 connection. It then starts forwarding the BCMCS flow.

7.3.4   BCMCS Bearer Release

The BCMCS bearer release procedure is responsible for tearing down the data path of a BCMCS bearer. The procedure can be triggered by several logical network nodes such as the BS or AN, by the PCF or by the BSN.

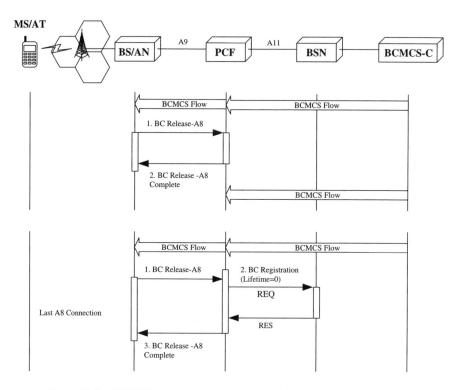

Figure 7.6   BCMCS bearer release procedure initiated by the MS or AT

Figure 7.6 illustrates the bearer release initiated by the BS or AN. The top part of the figure shows the case where the PCF keeps the A10 connection alive as it is still serving other BSs or ANs. The bottom part shows the case where the last BS or AN requests the release of the corresponding BCMCS bearer. The signalling steps of the BCMCS bearer release procedure are described in the following:

1. The BS or AT determines that the transmission of a BCMCS flow is no longer required, for example owing to the absence of BCMCS re-registrations. It initiates the release of the corresponding A8 connection by sending a *BC release-A8* message to its PCF.
2. If the PCF does not serve other BSs or ATs, it sends a *BC registration request* to its BSN, identifying the BCMCS flow ID in question and a lifetime attribute set to zero in order to indicate the release of the A10 connection. The BSN replies with a *BC registration response*, tearing down the A10 connection for the BCMCS flow.
3. If the PCF is serving other BSs or ANs for the BCMCS flow, then Step 2 is not required. Instead, the PCF replies with a *BC release-A8 complete* message to tear down the A8 connection. It then stops forwarding the BCMCS flow to the BS or AT in question.

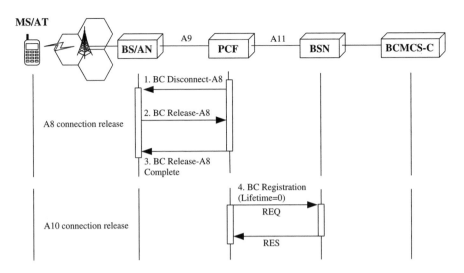

Figure 7.7   PCF-initiated BCMCS bearer release procedure

Figure 7.7 illustrates the PCF-initiated bearer release. The PCF-initiated bearer release consists of the following signalling steps:

1. When the PCF determines that an A8 connection is no longer required for a BCMCS flow, it initiates the release of the connection by sending a *BC disconnect-A8* message to the respective BS or AN.
2. The BS or AS responds to the request by sending a *BC release-A8* message to its PCF.
3. The PCF replies with a *BC release-A8 complete* message to tear down the A8 connection. It also stops forwarding the BCMCS flow.
4. The PCF may also determine that it does not require the BCMCS flow for any of the BSs or ANs it was serving and may trigger the release of the corresponding A10 connection. It does so by sending a *BC registration request* to its BSN, identifying the BCMCS flow ID in question. The lifetime attribute is set to zero to indicate the release of the A10 connection. The BSN replies with a *BC registration response*, tearing down the A10 connection for the BCMCS flow.

The BCMCS bearer release initiated by the BSN is depicted in Figure 7.8. The corresponding signalling flow is described in the following:

1. The BSN initiates the release of an A10 connection by sending a *BC registration update* to the PCF, providing the BCMCS flow ID in question. The PCF replies with a *BC registration acknowledge* message.
2. The PCF releases all existing A8 connections associated with the BCMCS flow. This is done by triggering the PCF-initiated bearer release described earlier.
3. The PCF tears down the A10 connection after all A8 connections have been released.

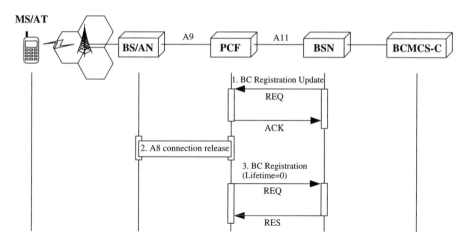

Figure 7.8 BSN-initiated BCMCS bearer release procedure

7.4 BCMSC Service Layer

The BCMCS service layer builds on the BCMCS bearer services to deliver multimedia broadcast content to mobile users. While the MBMS standard provides a detailed specification of the MBMS service layer, the BCMCS standard is much less explicit in defining its service layer. This may be attributed to the fact that the BCMCS service layer is much less complex than the MBMS service layer, with BCMCS only supporting streaming services and not reliable file delivery, as is the case with MBMS.

Services that make use of BCMCS capabilities are referred to as BCMCS broadcast programmes. Broadcast programmes are equivalent to MBMS user services. A broadcast programme is usually described by a programme name and a programme ID. Each programme provides a BCMCS content stream, consisting of one or more IP multicast data flows. These multicast data flows are commonly referred to as BCMCS flows in the standards. IP multicast flows are described by a destination IP multicast address, destination transport-layer port number, source IP address and port number. For efficient transport in the network, IP multicast flows are described by shorter BCMCS flow IDs. There is a one-to-many mapping between a programme name and BCMCS flow IDs. As an example, a football broadcast programme could offer an audio and a video stream as two separate BCMCS flows. A mobile user could either receive both flows as part of the programme or choose only to receive one of both flows. Both audio and video could also be encoded and transmitted as a single BCMCS flow. Each BCMCS flow is transmitted by means of a separate BCMCS bearer service. These bearer services can be provisioned either statically or dynamically.

Figure 7.9 shows the BCMCS protocol stack for delivery of BCMCS user data. Similarly to the MBMS protocol stack for streaming delivery, it makes use of widely deployed real-time multimedia streaming protocols of the IETF. These include the RTP and RCTP protocols that operate on top of the UDP and optionally the SRTP if the content is encrypted. Different media codecs are supported on top of the real-time multimedia protocols.

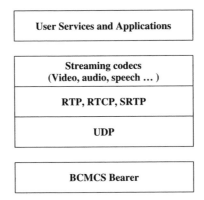

Figure 7.9   BCMCS user service protocol stack

This section describes the BCMCS information acquisition procedure, the BCMCS flow management procedure and the BCMCS security architecture. The BCMCS information acquisition procedure allows a mobile station to receive detailed information of a given BCMCS flow, whereas the BCMCS flow management procedure allows the content provider to configure BCMCS flows at the BCMCS controller. Finally, with the BCMCS security architecture, unauthorized access to BCMCS content is prevented.

7.4.1   BCMCS Information Acquisition

Service discovery in BCMCS provides the user with service-related information of broadcast programmes. These service announcements can carry information such as the programme name, a description of programme content and other service-related parameters such as the session start time or delivery conditions. In contrast to the session announcements in MBMS, service announcements in BCMCS do not provide sufficient information for the establishment of a BCMCS bearer service. This information is requested in a separate step with the BCMCS session acquisition procedure.

With the BCMCS information acquisition procedure, an MS or AT is able to obtain the parameters required to establish a BCMCS bearer service for one or more BCMCS broadcast programmes. These include parameters such as the BCMCS flow identity, BCMCS application information, BCMSC security parameters and/or link-layer information, depending on what encryption method is used for the session. During information acquisition, authentication and authorization of the user may also be performed.

Figure 7.10 shows the corresponding signalling flow of the BCMCS information acquisition procedure. The user typically connects to the network via a PTP bearer service to interact with the BCMCS controller and performs the procedure as follows:

1. The MS or AT initiates the procedure by sending a *BCMCS HTTP information* request message to the BCMCS-C over a unicast bearer service, providing the name of a broadcast programme.
2. The BCMCS issues an access request message to the Home RADIUS Server (HRS), provided that authentication and integrity protection is required, but neither an authorization header is presented in the received HTTP information request nor a valid nonce

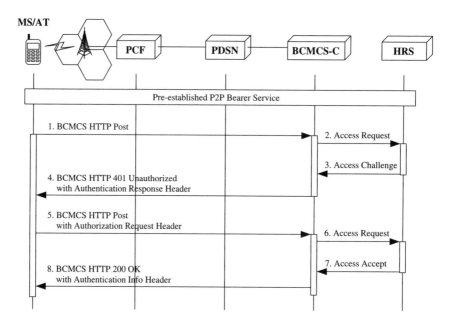

Figure 7.10   BCMCS information acquisition procedure

is available at the BCMCS-C. In security engineering, a nonce stands for *number used once*.

3. The HRS chooses a digest-nonce for the requesting user and responds with an access challenge back to the BCMCS-C, providing the chosen nonce and other digest attributes.

4. The BCMCS responds back with a *BCMCS HTTP 401 unauthorized* message to the MS or AT, providing a WWW-authentication header field with the digest challenge.

5. Upon receiving the challenge, the MS or AT computes the digest response, generates an WWW-authentication header containing the digest response and sends it as part of another *HTTP information request* message.

6. The BCMCS-C sends another access request to the HRS, this time with the digest response contained in the WWW-authentication header, and awaits authorization by the HRS.

7. Upon successful authentication, the HRS returns an *access accept* message providing the BCMCS with security-related parameters.

8. The BCMCS controller uses the provided information to authenticate and protect the response message and information contained in the message body. The response message returned to the MS or AT is a *BCMCS HTTP 200 OK* message with an authentication-info header, containing the BCMCS flow information, application information and relevant security information for accessing the BCMCS bearer service.

7.4.2 BCMCS Flow Management

The BCMCS standard also defines messages that allow the interaction of the BCMCS-C with the BCMCS Content Server (CS) in order to manage the provision of BCMCS

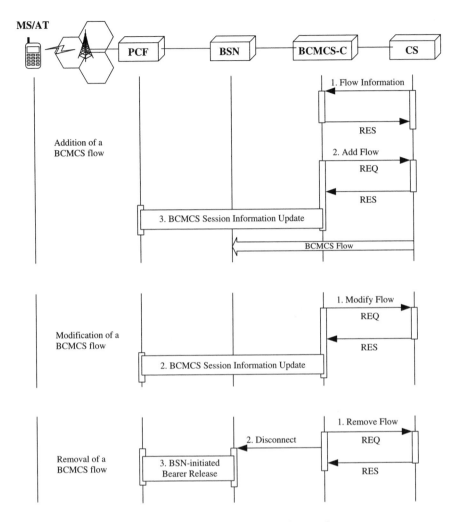

Figure 7.11   BCMCS flow control procedures

flows. Figure 7.11 depicts the different interactions. The *add flow* procedure allows the BCMCS-C to request resource reservation for BCMCS flows of upcoming sessions. The *modify flow* procedure can be used to modify parameters such as QoS or schedules of already added BCMCS flows. The *remove flow* procedure removes existing BCMCS flows from the BCMCS CS. The signalling flow of the *add flow* procedure is depicted in the upper part of Figure 7.11 and consists of the following steps:

1. Before initiating the add flow procedure, the CS has informed the BCMCS-C of BCMCS flows via the flow information message. The BCMCS-C confirms this with a *flow information* response.
2. At an appropriate time, the BCMCS-C performs a resource reservation at the CS for a BCMCS flow by sending an *add flow* request. The request provides the CS with a

detailed session schedule and other service-related parameters including a flow handle that the CS will use for future reference. A successful reservation is indicated by the *add flow* response.

3. On receipt of a successful *add flow* response, the BCMCS-C triggers the BCMCS session information update procedure. The CS starts sending the BCMCS flow to the next-hop multicast router towards the BSN at the scheduled transmission time.

Start and end times of a previously added flow can be modified by the *modify flow* procedure. The *modify flow* can be used to extend the session time of an ongoing BCMCS flow, which is depicted in the middle of Figure 7.11 and consists of the following steps:

1. The BCMCS-C sends a *modify flow* request message to the CS, identifying the BCMCS flow in question by the previously provided flow handle and specifying the new end time for the BCMCS flow. After modification, the CS indicates the success of the modification procedure with a *modify flow* response.
2. Upon success, the BCMCS-C triggers BCMCS session update procedure in order to configure the bearer service according to the new parameters.

Finally, the removal of a reservation of a previously added BCMCS flow is accomplished with the *remove flow* procedure. The lower part of Figure 7.11 illustrates the resulting signalling flow for the case where a currently provisioned BCMCS is removed. The steps for the *remove flow* procedure are described as follows:

1. The BCMCS-C sends a *remove flow* request message to the CS, identifying the BCMCS flow in question by the previously provided flow handle. The CS indicates the successful removal of the flow with the *remove flow* response.
2. Meanwhile, the BCMCS-C triggers the release of the bearer plane for the BCMCS flow by sending a RADIUS disconnect message to the BSN.
3. The BSN stops forwarding content for the identified BCMCS flow and triggers the release of the BCMCS bearer plane.

7.4.3 BCMCS Security

The BCMCS service layer allows content providers to offer content as BCMCS programmes, either free of charge or based on user subscriptions. BCMCS programmes that do not require subscription are offered for free and can be received by any mobile user located in the service area. Subscription-based BCMCS programmes, on the other hand, allow content providers to charge a subscription fee and should consequently only be received by valid subscribers. The purpose of BCMCS security is to prevent unauthorized users who are not subscribed to a BCMCS programme from gaining access to the transmitted BCMCS flows of the programme. Preventing unauthorized access is achieved by encryption of the BCMCS flows prior to transmission. Besides encryption, BCMCS security functions have to ensure that only valid subscribers have the correct encryption key to decrypt the received BCMCS flows and gain access to the transmitted content.

Figure 7.12 provides an overview of the BCMCS security architecture. As shown in the figure, the BCMCS content is encrypted by means of the SK prior to transmission and

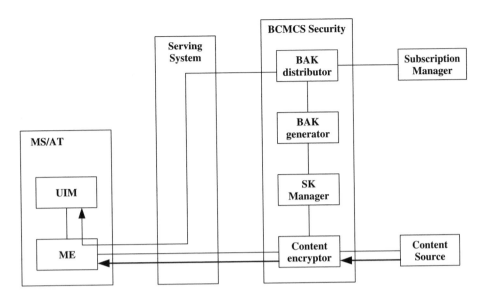

Figure 7.12 BCMCS security architecture

decoded at the receiver with the same SK. The SK is changed frequently in order to avoid a rogue subscriber from sharing its key with unauthorized users. The SK is never directly transmitted but rather derived from the BAK and a random value transmitted alongside the encrypted content.

The BAK provides access to one or more BCMCS flows of a BCMCS broadcast programme for a limited duration of time. As the BAK is used to derive the SK, its validity timeframe is usually longer. BAKs are provisioned into the secure memory of the User Identity Module (UIM) of a valid subscriber via the BAK distributor. In order to avoid having unauthorized users gain access to BAKs from BAK distributors, for example by falsely providing the identity of a subscribed UIM, BAKs are encrypted prior to their distribution. For this encryption, the BAK distributor uses a Temporary Key (TK), which is computed from a Registration Key (RK), which is only shared between valid subscribers of a BMCMS programme and the corresponding BCMCS subscription manager.

7.5 Summary

This chapter has provided a detailed description of the protocols and mechanisms used in BCMCS to deliver multicast and broadcast services to mobile users. The BCMCS service framework can be split into the network and service layer. The BCMCS network layer provides the BCMCS bearer services for efficient delivery of BCMCS flows, as well as the necessary service functions for managing the BCMCS bearer services. The BCMCS bearer path architecture can be split into two levels, namely the IP multicast tree from the BCMCS content server up to the BSN and the link-level BCMCS tree from the BSN to the receivers. Several protocol enhancements have been made to existing interfaces within the radio access and core network in order to accommodate BCMCS bearer functionality. The management of the bearer paths is handled by bearer service procedures, which are part of the bearer service control plane. Bearer service procedures operate in a distributed fashion over the different logical network nodes along the bearer path and are responsible for the establishment, maintenance and release of data plane bearers. Several important procedures were presented, with detailed signalling flows between the different network entities.

The BCMCS service layer consists of the functions required for the set-up and management of BMCMS broadcast sessions. The protocol stack used for the delivery of BCMCS programmes was presented, alongside a description of important BCMCS service procedures such as BCMCS information aquisition and BCMCS flow management.

8

Multicast Capacity over the CDMA Air Interface[†]

8.1 Introduction

In CDMA, data transfer to a group may either take place on multiple PTP channels transmitted to individual multicast users separately or on a single PTM channel that is broadcast over the entire cell. The trade-off between transmitting multicast data to individual users in a unicast fashion and broadcasting multicast data over the entire cell is complex owing to the wide range of radio-specific parameters that must be taken into account. For instance, macrodiversity gain by virtue of soft handover is possible with PTP channels, whereas PTM channels only support hard handover owing to the distributed nature of the users on the PTM channel. PTP channels, however, are less efficient in terms of code usage, since each individual PTP channel requires a separate spreading code. On the other hand, broadcasting multicast data over the entire cell utilizes a single spreading code.

In this chapter, we analyse the capacity of both channel selection schemes by deriving closed-form analytical expressions for the Erlang capacity of both PTP and PTM channels. Our analytical study takes into account deterministic path loss, lognormal shadowing, code orthogonality, power control error and the impact of hard and soft handover.

Based on the closed-form expressions for the PTP and PTM channels, we perform a comparative analysis of the capacity for both approaches and also investigate the impact of important radio-specific parameters such as the level of orthogonality and signal-to-interference ratio requirements, as well as system-wide parameters such as the multicast activity factor and power control error. The chapter provides insight into the trade-offs between employing PTP and PTM channels for multicast over the CDMA air interface.

[†] Based on "The Capacity of Different Channel Selection Strategies for Multicast Transmissions in WCDMA", Rümmler, R.; Ashraf, I.; Aghvami, A. H.; Which appeared in IEEE Transactions on Vehicular Technology, Volume 56, Issue 4, Part 2, July 2007, page(s): 2180–2193.

8.2 PTP and PTM Channels for Multicast

In CDMA, data transfer to a group may take place on a single PTM channel or multiple PTP channels. This is illustrated in Figure 8.1. With the PTM scheme, a single channel is used to broadcast over the entire cell, whereas, with the PTP scheme, multicast traffic is duplicated within the radio access network for each user such that each user receives multicast traffic on a dedicated channel.

Several trade-offs exist between employing PTP or PTM channels for multicast over the CDMA air interface. In terms of code usage, PTP channels are less efficient since each individual PTP channel requires a separate spreading code. On the other hand, employing a PTM channel to broadcast multicast data over the entire cell utilizes a single spreading code. As CDMA systems are not code limited, this difference is not of critical importance. From a radio perspective, power control and soft handover are critical features of CDMA systems that have a significant impact on overall system capacity. We briefly describe the differences in how PTP and PTM channels support these features.

8.2.1 Power Control

CDMA is an interference-limited multiple-access system. With universal frequency reuse, all radio connections transmit on the same frequency. The goal of power control is to reduce the transmit power for each radio connection to a minimum while at the same time satisfying the SIR requirement of the radio connection. This is done in order to limit the interference seen by other radio connections.

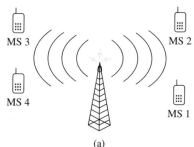

(a)
Multicast with single point-to-multipoint channel

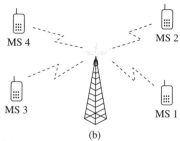

(b)
Multicast with multiple point-to-point channels

Figure 8.1 Channel selection for multicast over the CDMA air interface

With PTP channels, tight and fast power control can be achieved by means of open-loop and closed-loop power control. With PTM channels, however, owing to the one-to-many scenario, power control is more difficult. Conceptually, the BS must always transmit at a power that is sufficient for the MS that requires the most power. The BS must evaluate the power requirements for each multicast user in the group and transmit at a power that satisfies the worst-case power requirement. Doing so in a fashion similar to closed-loop power control in WCMDA, which executes power control commands at a rate of 1500 times per second, or 1.5 kHz, is difficult to achieve for large multicast groups as this would necessitate significant processing capacity at the BS. If power control is not feasible for PTM channels, providing a reliable radio link can be achieved by satisfying the SIR requirement for the worst case of a user located at the cell edge.

8.2.2 Soft and Hard Handover

Soft handover is a CDMA-specific feature that allows an MS to be simultaneously connected to two or more BSs during a radio connection. Soft handover results in a soft handover diversity gain. Soft handover is possible with PTP channels, whereas PTM channels only support hard handover owing to the distributed nature of the users on the PTM channel.

A hard handover takes place in a break-before-make fashion, that is, the channel in the source cell is released and only then is the channel in the target cell engaged. Thus, the connection to the source is broken before the connection to the target is made. Hard handovers are intended to be instantaneous in order to minimize the disruption to the radio connection. With PTM channels that only support hard handover, the BS needs to transmit at a higher power level that can be received beyond the cell boundary in order to ensure that the hard handover can take place smoothly. Overall, this means that the PTM channel requires more power than a comparable radio connection that supports soft handover.

8.3 System Model

Having briefly touched upon the main system-wide differences in employing PTP and PTM channels for multicast, we now describe our system model.

We consider a DS-CDMA system. In CDMA systems, as in other cellular systems, a service area is divided into cells. A user's signal is spread over the entire allocated bandwidth. Owing to the nature of spread-spectrum systems such as CDMA, the same radio frequencies can be used in each cell. This is referred to as universal frequency reuse. The separation of different users' signals is achieved by means of signature sequences used to spread the spectrum. In WCDMA, for instance, the data for all users are spread by OVSF codes. The spread signals of all users are combined synchronously, and prior to transmission the sum is scrambled by a long pseudo-noise code that is unique to a particular BS.

The service area of our system is divided into $K_c = 19$ hexagonal cells of equal size arranged in three tiers, with a BS located at the centre of each cell. Each BS is equipped with an omnidirectional antenna. The radius for each cell, denoted by R_c, is normalized to unity. Each MS is assumed to be located at random, independently of all the other users, and uniformly distributed over the service area.

8.3.1 Propagation Model

For the propagation model, we firstly consider a deterministic dependence of the received power on the distance between transmitter and receiver. With d denoting the distance between transmitter and receiver, the average received power is proportional to $d^{-\omega}$, where ω is the path loss exponent, typically 3–4. Terrain irregularities give rise to another source of attenuation, shadow fading. In a popular model, based on available measurements, this random effect is described by a lognormal random variable (Zorzi, 1996).

The propagation loss due to path loss and shadow fading between an MS and the jth BS is thus given by

$$L_j = d_j^{-\omega} \cdot 10^{\xi_j/10} = d_j^{-\omega} \cdot \chi_j,$$

where d_j is the distance between the MS and the jth BS, ω is the path loss exponent, ξ_j is a Gaussian distributed random variable with zero mean and standard deviation σ representing shadow fading and χ_j is a lognormal random variable that is equal to $10^{\xi_j/10}$. The standard deviation σ has units of decibels, and is referred to as the shadowing or decibel spread, with values ranging between 4–12 dB depending on the severity of the shadow fading channel (Slimane, 2001).

In CDMA systems, the receiver is able to combat the effect of fast fading if there are more than two multiple paths that can be resolved by the receiver. By using a Rake receiver incorporated with closed-loop power control, bit interleaving, channel coding and space diversity reception, the effect of fast fading can be reduced to a minimum (Gilhousen et al., 1991; Simpson and Holtzman, 1993). Therefore, a radio channel with only path loss and slow fading is considered in our analysis on the presumption that any fast fading would be effectively combated by the receiver (Lee and Steele, 1998).

8.3.2 Interference Model

On the downlink, each BS transmits to its own users, using direct-sequence spread-spectrum modulation. Each MS, along with the desired signal, picks up interfering signals. In general, these signals come from both the BSs of the surrounding cells (intercell interference) and from the BS of the cell in which the user is located (intracell interference). The interference power is the difference between the total received power and the desired signal power. Since the intracell signals are all transmitted by the same BS, they can be synchronized. In the situation of perfect orthogonality of the spreading sequences, the intracell interference is practically reduced to zero. However, perfect orthogonality is usually not the case owing to multipath propagation.

The intracell interference power at the ith MS is given by

$$I_{sc} = \varphi \cdot P_0 \cdot L_0, \tag{8.1}$$

where P_0 is the total transmitted signal power from the home BS, L_0 is the path loss from the home BS to the ith MS and φ is the orthogonality factor which represents the fraction of the received power on the downlink that is seen as interference by each of the codes from the same BS. The orthogonality factor φ accounts for imperfections in orthogonality among signature sequences owing to multipath propagation. If $\varphi = 0$, perfect orthogonality between signature sequences is maintained and $I_{sc} = 0$. For severe multipath, φ approaches unity.

The intercell interference power from adjacent BS to the ith MS is given by

$$I_{oc} = \sum_{k=1}^{K_c} P_k \cdot L_k, \qquad (8.2)$$

where P_k is the total transmitted power from the kth BS and L_k is the path loss from the kth BS to the ith MS. The transmitted signal power from the BS is actually a function of the number of ongoing sessions served by the BS and thus can be modelled as a random variable. We assume that all cells are uniformly loaded and statistically identical in terms of cell loading. As a result, we assume that all BSs are accommodating the maximum number of ongoing sessions and therefore transmitting at full power (Choi and Kim, 2001b). This corresponds to the worst-case condition. The worst-case condition is a popular approach that is typically used in cell planning (Jansen and Prasad, 1995).

We evaluate the performance of both multicast schemes by first deriving suitable outage probability expressions and then numerically evaluating the system capacity, in other words the number of users that can be accommodated in the system for a given required outage probability, which must be guaranteed for each subscriber in the network (Slimane, 2001; Zorzi, 1995).

8.4 Analysis of Multicast Capacity

We now derive the downlink capacity for the PTP and PTM channel selection schemes. Without loss of generality, we consider a single multicast group, of which K_m multicast group members are located within the reference cell.

Power control and site diversity by means of soft handover are fundamental in comparing the performance of both schemes for multicast transmissions. The aim of power control is to minimize the users' transmitted powers while maintaining a certain QoS for each user (Ulukus and Yates, 1998). Power control is considered to be the single most important system requirement for CDMA (Gilhousen et al., 1991). Power control in the uplink is used to combat the near-far problem by maintaining the received powers from all MS within a cell at a constant level and to reduce power consumption at the MSs (Zander, 1993). In the downlink, power control is mainly used to reduce the interference outside the boundaries of a single cell (Lee, 1991b) and is therefore not as critical to system performance as power control on the uplink (Viterbi, Gilhousen and Zehavi, 1994).

With soft handover, an MS can be connected to several BSs simultaneously, transmitting and receiving signals to and from more than one BS (Kim et al., 2002), thus resulting in site diversity gain. As with power control, site diversity by means of soft handover in the downlink is not as critical to system capacity as in the uplink. Soft handover provides a considerable gain in uplink performance with virtually no interference drawbacks. This is due to the fact that any BS engaged in soft handover would be receiving the signal from the MS in any case, without advantage if it were not in soft handover (Viterbi, 1995). In the downlink, however, two or more BSs involved in the soft handover connection must transmit at a given power level to the same MS, thus increasing the interference seen by other MSs. For the overall system, the increase in interference to MSs can outweigh the site diversity gain from soft handover (Kim *et al.*, 2002).

The total received power for a single MS from the home BS is $P_0 \cdot L_0$. Let ψ denote the fraction of total transmission power devoted to traffic sources, with $1 - \psi$ denoting the fraction of transmission power devoted to the pilot signal, as well as to any common information destined for all users (Viterbi, 1995). Let ϕ_i denote the fraction of transmission power allocated for the ith MS, then $\psi \cdot \phi_i \cdot P_0 \cdot L_0$ is the desired received power from the home BS. From Equation (8.1) and Equation (8.2), the received E_b/I_0 at the MS, denoted by γ, is given by (Choi and Kim, 2001a)

$$\gamma = \frac{W}{R} \cdot \frac{\psi \cdot \phi_i \cdot P_0 \cdot L_0}{\displaystyle\sum_{k=1}^{K_c} P_k \cdot L_k + \varphi \cdot P_0 \cdot L_0 + N_0 W},$$

where N_0 is the power spectral density of the background noise, W is the spreading bandwidth and R is the data rate.

CDMA systems are typically limited by interference (Zorzi and Milstein, 1994), and the background noise power is therefore generally considered to be negligible compared with the total signal power received from all BSs (Choi and Kim, 2001a). With $S_0 = P_0 \cdot L_0$, the received E_b/I_0 at the MS is

$$\gamma \approx \frac{W \cdot \psi}{R} \cdot \frac{\phi_i \cdot S_0}{I_{oc} + S_0 \cdot \varphi} \approx \frac{W \cdot \psi}{R} \cdot \frac{\phi_i}{\dfrac{I_{oc}}{S_0} + \varphi}. \tag{8.3}$$

In order to achieve the required performance, the E_b/I_0 received at the MS should be greater or equal to the required E_b/I_0. The aim of a power control algorithm is to minimize the transmission powers of the MSs or BSs while maintaining a certain QoS for each user. Without power control on the downlink, the BS transmits the same amount of power to all MSs. However, those MSs that are near the BS receive a stronger signal from their own BS and less interference from other BSs. By reducing the power transmitted to these MSs, yet satisfying the SIR requirements, the overall capacity can be increased (Alavi and Nettleton, 1982; Gejji, 1992; Lee, 1991a). With accurate power control, the performance is independent of the user's location (Zorzi, 1996). From Equation (8.3), the fraction of power from the home BS allocated to the ith user, assuming perfect power control, is

$$\phi_i = \frac{\gamma \cdot R}{W \psi} \cdot \left(\frac{I_{oc}}{S_0} + \varphi \right). \tag{8.4}$$

It is evident from Equation (8.4) that the fraction of total BS transmit power allocated to the ith user is proportional to I_{oc}/S_0, the ratio of intercell interference power to received power from the home BS. However, I_{oc}/S_0 is dependent on the location of the ith user. We therefore write $z_i(r_i, \theta_i) = I_{oc}/S_0$, where (r_i, θ_i) represents the polar coordinate of the ith MS. Equation (8.4) may then be rewritten as

$$\phi_i = \frac{\gamma \cdot R}{W \psi} \cdot [z_i(r_i, \theta_i) + \varphi] = \frac{\gamma \cdot R}{W \psi} \cdot y_i(r_i, \theta_i).$$

Since $S_0 = P_0 \cdot L_0$ is a lognormal random variable and I_{oc}, the sum of 18 lognormal random variables defined according to Equation (8.2), may be approximated as a lognormal random variable (Cardieri and Rappaport, 2001), $z_i(r_i, \theta_i) = I_{oc}/S_0$ is approximated by a lognormal random variable with mean m_z and standard deviation σ_z, both in dB. Similarly, $y_i(r_i, \theta_i) = z_i(r_i, \theta_i) + \varphi$ is also a lognormal random variable with mean m_y and standard deviation σ_y, both in dB. In order to get a handle on the worst-case I_{oc}/S_0, we calculate I_{oc}/S_0 exactly at the cell boundary and only consider the interference from the 11 closest and therefore dominant BSs. This is done in Equation (A.1) of Appendix A.

8.4.1 Multicast Capacity with PTP Channels

Soft handover results in macrodiversity gain. We consider the impact of macrodiversity gain quantitatively in terms of increased interference power and qualitatively in terms of improved handover performance, resulting in lower handover latency and lower probability of packet loss during handover. Handover latency and probability of packet loss, the latter usually occurring as a result of the former, are particularly important in the context of multicast, since reliable transmission to a group is difficult to achieve, thus acting as a bottleneck in the wider acceptance of multicast at the present time.

With soft handover, the highest transmit power is required when the MS is located exactly at the cell boundary (Viterbi, 1995). The maximum fraction of BS transmit power for a user on a PTP channel is

$$\phi_{\max} = \frac{\gamma \cdot R}{W \psi} \cdot [z(R_c) + \varphi] = \frac{\gamma \cdot R}{W \psi} \cdot y(R_c).$$

With power control, the fraction of power allocated to individual users is directly proportional to $z_i(r_i, \theta_i) = I_{oc}/S_0$. We are interested in the variation of $z_i(r_i, \theta_i)$ under the assumption that all users are uniformly distributed over the cell area. The variation of I_{oc}/S_0 within the cell is only a function of the distance-dependent path loss, which is entirely deterministic. This allows us to analyse the variation of I_{oc}/S_0 over the cell area by means of numerical integration. More specifically, we are interested in evaluating the ratio between I_{oc}/S_0 averaged over the entire cell area and the worst-case I_{oc}/S_0 averaged over the cell boundary. This allows us to express the average required downlink power for an individual user with soft handover in terms of the worst-case required power at the cell boundary (Choi and Kim, 2001a; Lee and Miller, 1999; Viterbi, 1995). This is convenient, since the worst-case I_{oc}/S_0 incorporating the effect of shadowing at the cell boundary may be calculated using Equation (A.1).

The average required downlink power for an individual user with soft handover, expressing the variation of I_{oc}/S_0 in relation to the worst-case required power at the cell boundary (Choi and Kim, 2001a; Lee and Miller, 1999; Viterbi, 1995), is denoted by η and derived in Section A.2 of Appendix A.

In evaluating the capacity of the PTP channel selection scheme, we first derive the BS outage probability, i.e. the probability that the total BS power is not sufficient to satisfy the power requirements of all users. The BS outage probability in a multiservice environment consisting of K_v voice users and K_m data users is denoted by P_b. The Erlang capacity of the PTP channel selection scheme is then defined as the set of values of A_v and A_m that keep the BS outage probability P_b at a target level, typically 0.01 (Choi and Kim, 2001a). The Erlang capacity is derived in Section A.3 of Appendix A.

8.4.2 Multicast Capacity with PTM Channels

We consider fixed power allocation for the stand-alone common or PTM channel carrying multicast traffic. As such, the power allocated to the PTM channel for multicast is determined by downlink transmit power and cell coverage planning (Ishikawa, Hayashi and Onoe, 2002). We investigate the transmit power needed in order to satisfy a QoS target and coverage requirements imposed by hard handover. Choosing a fairly high QoS target for the PTM channel effectively introduces an additional fading margin in order to overcome losses resulting from small-scale fading.

Owing to the distributed nature of users receiving the PTM channel, soft handover is not possible with PTM channels. With hard handover, the ideal condition of handover at the cell boundary is both unrealistic and undesirable. This is because it can lead to the *ping-pong* effect, where a user near the boundary is handed back and forth several times from one BS to the other. In practical hard handover systems, the handover occurs only after the first cell's BS power is reduced far enough below its value at the boundary (Viterbi, Gilhousen and Zehavi, 1994). This is illustrated in Figure 8.2, which depicts the connection probability at a distance d from the BS for different values of shadowing spread. When no shadowing is considered, the connection probability beyond the cell boundary (in other words, $d > 1.0$) is zero. With shadowing, however, the connection probability at a distance $d = 1.3$ with $\sigma = 4$ dB is approx. 0.1, whereas for $\sigma = 8$ dB a connection probability of 0.1 is observed at $d = 1.5$.

Two methods for macrodiversity, namely selective combining and soft combining, are supported in MBMS. With selective combining, a UE receives and simultaneously decodes packets from radio links of multiple adjacent cells/sectors. The received packets are then handled at the RLC layer. The RLC layer keeps the first correctly received packet and discards all other packets. A packet that is erroneously received on one link may be easily recovered by a correctly received packet from another link. In contrast, soft combining mitigates the reception of an erroneous packet on the physical layer. Incoming packets

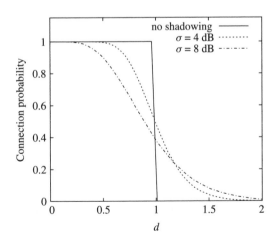

Figure 8.2 Connection probability against normalized distance d from home BS for different values of shadowing spread

are directly combined on the physical layer and decoded after combination. The effect of selective and soft combining are not considered in our analysis.

In evaluating the capacity of the PTM scheme, we first define the link outage probability as the fraction of time that the fixed fraction of power allocated to the PTM channel is not sufficient to provide an MS at a distance $r_c > R_c$ from the home BS. The link outage probability, denoted by P_c, is used to calculate the fixed fraction of power allocated to the PTM channel, denoted by ϕ_c, such that the target link outage probability P_c is satisfied. The capacity of the PTM scheme is derived in Section A.4 of Appendix A.

8.5   Numerical Results

In this section, we present performance results of both channel selection strategies for multicast in CDMA. For our numerical evaluation, we rely on system parameters that are specific to WCDMA. Extending the results to other CDMA air interfaces such as the air interface of CDMA2000 is trivial and left for further study.

We perform both a comparative performance study, which gives insight into the trade-offs between each of the two schemes, and a sensitivity analysis to different system and propagation parameters. We firstly investigate the capacity of each scheme in a single-service environment consisting solely of multicast users. A convenient performance metric in a single-service environment is the fraction of BS power required to support a given amount of multicast traffic. We then show the achievable capacity in a mixed-traffic environment, in which multicast users coexist with regular voice users. In the mixed-traffic environment, the performance metric is the voice capacity that can be supported alongside a given amount of multicast traffic.

For our comparative performance study, we choose $\omega = 4$ as the path loss exponent (Jakes, 1994), $\psi = 0.8$ as the fraction of BS power assigned to traffic channels (Gilhousen *et al.*, 1991) and $\sigma = 4.5$ dB as the shadowing spread (Zorzi, Rossi and Mazzini, 2002). The duty cycle for voice is chosen as $\alpha_v = 0.375$, which is a typical value reported by Gilhousen *et al.* (1991). We assume that a fraction $g < 1$ of all users are in a two-way soft handover, and that both BSs involved in the soft handover allocate essentially the same fraction of power to the soft-handover users. The fraction of total users in soft handover is taken as $g = 0.25$ (Viterbi, 1995). We take $R_v = 30$ kbps as the physical-layer bit rate for voice, and $R_m = 60$ kbps as the physical-layer bit rate for multicast. It should be noted that R_v and R_m are the respective bit rates after coding, with corresponding spreading factors of $W/R_v = 128$ for voice and $W/R_m = 64$ for multicast respectively. Using a half-rate convolutional encoder, the useful bit rate for multicast is then approximately 30 kbps.

The bit energy-to-interference requirement γ_m depends on the system operating conditions, including vehicular speed and link-level parameters (Choi and Kim, 2001a). With soft handover, γ_m is somewhat lower compared with the case where only hard handover can be performed. We neglect this additional advantage in E_b/I_0 requirement between hard and soft handover since it has little impact on performance (Viterbi, Gilhousen and Zehavi, 1994). In the results presented here, the average value of required E_b/I_0 for multicast is taken as $\gamma_m = 3$ dB, both on PTP and on PTM channels. For voice, $\gamma_v = 4$ dB is chosen as the required E_b/I_0 (Choi and Kim, 2001a). We investigate the sensitivity of both multicast schemes with respect to the activity factor of the multicast source, α_m, the

Table 8.1 System parameters

Parameter	Variable	Value
Path loss exponent	ω	4
Shadowing spread	σ	4.5 dB
Fraction of total power for user traffic	ψ	0.8
BS outage probability	P_b	0.01
Spread of PCE	σ_e	0.5 dB
Downlink orthogonality	φ	0.5
Fraction of users in soft handover	g	0.25
Duty cycle of voice traffic	α_v	0.375
Duty cycle of multicast traffic	α_m	0.5
Required E_b/I_0 of voice sources	γ_v	4 dB
Required E_b/I_0 of multicast sources	γ_m	3 dB
Voice bit rate	R_v	30 kbps
Multicast bit rate	R_m	60 kbps

level of orthogonality present on the downlink, φ, and the required E_b/I_0 of the multicast source, γ_m. We also consider the spread of the Power Control Error (PCE). When power control is not ideal, the error is assumed to be lognormally distributed with standard deviation σ_e in dB (Tam and Lau, 1999). In comparing the relative performance differences between the two schemes, we choose the following nominal values for the system parameters: $P_b = 0.01$ as the target BS outage probability, $\alpha_m = 0.5$ as the activity factor of the multicast source, $\varphi = 0.5$ as the orthogonality factor and $\sigma_e = 0.5$ dB as the PCE spread. Table 8.1 summarizes the system parameters used in assessing the performance of each scheme.

8.5.1 Comparative Analysis

We now present results of the comparative performance analysis, which highlights the trade-off between the PTP and PTM schemes in both a single-service and multiservice traffic environment.

PTP Channel Selection

We firstly consider the fraction of transmit power required to support a given amount of multicast traffic with PTP channels. Figure 8.3 plots the supported Erlang capacity for multicast traffic, $A_m = \lambda_m/\mu_m$, as a function of the fraction of total BS power assigned to the multicast source, ϕ_m, for different values of the BS outage probability, $P_b = 0.01$, 0.02, 0.03. As can be seen from the figure, a decrease of P_b from 0.03 to 0.01 leads to a capacity reduction of approx. 25 %. Taking $P_b = 0.01$, $A_m = 9$ Erlang can be supported if the total BS transmission power is allocated for multicast.

We now illustrate the performance of the PTP scheme in a traffic environment consisting of voice and multicast users. Figure 8.4 illustrates the Erlang capacity for voice, A_v, against the Erlang capacity for multicast, A_m. The curves in Figure 8.4, which are plotted for $P_b = 0.01$, 0.02, 0.03 and 0.05, represent the set of values for A_v and A_m that keep the BS outage probability at a required target level. We observe that an increase in P_b from

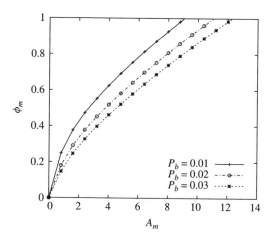

Figure 8.3 ϕ_m against A_m for different values of P_b with PTP scheme

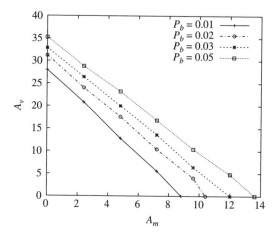

Figure 8.4 A_v against A_m for different values of P_b with PTP scheme

0.01 to 0.02 results in a capacity increase of approx. 15%, with the margin remaining constant over the entire range of the graph. The difference in performance between $P_b = 0.02$ and 0.03 is marginally smaller compared with the difference between $P_b = 0.01$ and 0.02 and amounts to approximately 10%. Similarly, the difference in capacity between $P_b = 0.03$ and 0.05 is also in the range of 10%. For the system parameters given in Table 8.1, and taking $A_v = 15$ as a reference case, a multicast capacity of $A_m = 4.5$ can be accommodated.

PTM Channel Selection

With the PTM channel selection scheme, a fixed amount of power is assigned to the PTM channel carrying multicast traffic. As such, we firstly investigate the power that is

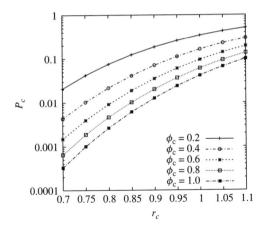

Figure 8.5 P_c against r_c for different values of ϕ_c with PTM scheme

needed in order to satisfy a required link outage probability, P_c, at a distance r_c from the home BS. We then investigate how much voice capacity can be accommodated alongside a PTM channel that is allocated a given fraction of transmit power.

In order to illustrate the coverage that may be achieved with a PTM channel that is allocated a fraction of transmit power ϕ_c, Figure 8.5 plots P_c as a function of r_c for different values of ϕ_c. As described earlier, hard handover typically occurs when the user has moved a reasonable distance beyond the cell boundary. The fraction of transmit power for the PTM channel, ϕ_c, must therefore be sufficient to provide suitable performance at a distance $r_c > R_c$ from the home BS.

It can be seen from Figure 8.5 that, in order to achieve $P_c = 0.1$ at $r_c = 1.02$, $\phi_c = 0.6$ is required, that is, 60 % of total BS transmit power must be reserved for the PTM channel. The fraction of power for the PTM channel increases accordingly when more coverage is needed in order to achieve suitable handover performance. Fractionally better performance of $P_c = 0.07$ at the very edge of a possible handover may be achieved with $\phi_c = 0.8$.

Figure 8.6 plots a similar result to Figure 8.5, but from a different perspective. Since we are interested in the achievable link outage probability at $r_c > R_c$, in other words beyond the cell boundary, Figure 8.6 plots P_c against ϕ_c for different values of r_c. For small values of ϕ_c, there is hardly any difference in P_c for the different values of r_c. However, this changes as P_c increases. As one might expect, increasing r_c results in a decrease in the corresponding value of P_c.

The power allocated to the PTM channel is only broadcast by the BS when there is data to transmit. Taking into account the duty cycle of the multicast source, which was not considered for the previous figures, we can now investigate the performance of the PTM scheme in a mixed-traffic environment consisting of voice and multicast users.

Figure 8.7 plots A_v against ϕ_c for different values of P_b, the BS outage probability, taking $\alpha_m = 0.5$. Clearly, increasing the BS outage probability requirement for voice results in a decreased capacity. Also, since the multicast source is not transmitting continuously, an Erlang capacity of $A_v = 13$ can be supported even if $\phi_c = 1$.

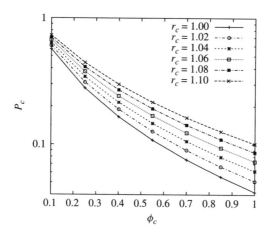

Figure 8.6 P_c against ϕ_c for different values of r_c with PTM scheme

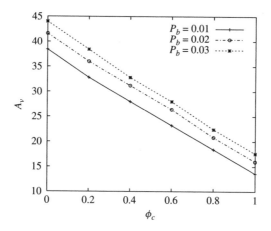

Figure 8.7 A_v against ϕ_c for different values of P_b with PTM scheme

8.5.2   Sensitivity Analysis

We now present results of the sensitivity analysis of the three schemes towards a number of system and propagation parameters, such as the multicast activity factor, the orthogonality factor and the standard deviation of the power control error, as well as the E_b/I_0 requirement of the multicast service.

Multicast Activity Factor

The multicast activity factor is a measure for how much of the time the source transmits throughout the session. With low-activity sources, the multicast activity factor is close to zero, whereas, with sources that are transmitting continously, the multicast activity

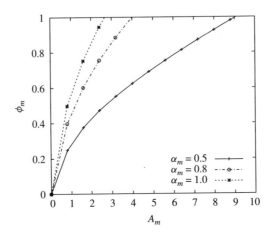

Figure 8.8 ϕ_m against A_m for different values of α_m with PTP scheme

factor is close to 1. We firstly investigate the impact of the multicast activity factor on the capacity of both schemes. Figure 8.8 plots ϕ_m against A_m for the PTP scheme, with $\alpha_m = 0.5$, 0.8 and 1.0. For simplicity, we assume that the activity factor of individual multicast users is equal to that of the activity factor of the multicast source. This is a reasonable assumption if only a small amount of retransmissions are performed over the link layer by means of ARQ. As can be seen from Figure 8.8, α_m has a large impact on capacity. The maximum Erlang capacity with $\alpha_m = 0.5$ is $A_m = 9$, whereas with $\alpha_m = 0.8$ the capacity is $A_m = 4$. This amounts to a reduction in capacity by a factor of more than 2. A similar trend can be observed between $\alpha_m = 0.8$ and $\alpha_m = 1.0$, for which the capacity is reduced by a factor of 1.5. When the multicast source is transmitting continuously and $\alpha_m = 1.0$, the maximum Erlang capacity is $A_m = 2.7$. Figure 8.9, which plots A_v against A_m for $\alpha_m = 0.6$, 0.8 and 1.0, confirms the result from the previous figure, in that the multicast activity factor has a large impact on system performance.

Figure 8.10 plots the voice capacity in Erlang, A_v, against the fraction of BS transmit power allocated to the multicast source, ϕ_c, for the PTM scheme. Figure 8.10 is plotted for $\alpha_m = 0.5$, 0.8 and 1.0. As expected, for $\alpha_m = 1.0$, the voice capacity tends towards zero as ϕ_c approaches unity because all available power is used for multicast. The activity factor of the multicast source is therefore an important factor to be considered in reserving a fixed fraction of transmit power for the PTM channel, since the activity factor of the multicast source will determine how much additional voice capacity can be accommodated in the system.

Level of Orthogonality

We now investigate the impact of the downlink orthogonality factor on the performance of the three multicast schemes. Figure 8.11 plots ϕ_m against A_m for the PTP scheme, with $\varphi = 0.0$, 0.5 and 1.0. A value of $\varphi = 0.0$ corresponds to perfect orthogonality, whereas $\varphi = 1.0$ is representative of a complete loss of orthogonality. As can be seen from Figure 8.11, a complete loss of orthogonality has only a negligible impact on performance compared with perfect orthogonality.

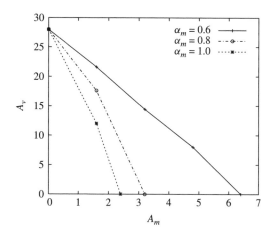

Figure 8.9 A_v against A_m for different values of α_m with PTP scheme

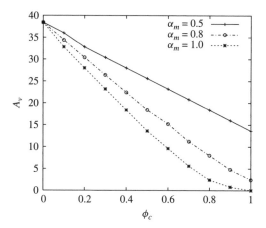

Figure 8.10 A_v against ϕ_c for different values of α_m with PTM scheme

Whereas the PTM channel is not affected by PCE since power allocation is fixed, Figure 8.12 illustrates the impact of loss of orthogonality on the coverage of the PTM channel by considering propagation with complete orthogonality and no orthogonality for $r_c = 1.01$ and 1.05. Figure 8.12 shows that the performance of the PTM channel is sensitive to the level of orthogonality present on the downlink. In particular for $\phi_c \leq 0.25$, the link outage probability for $r_c = 1.05$ with $\varphi = 0.0$ is equal to the link outage probability for $r_c = 1.01$ with $\varphi = 1.0$.

Power Control Error

We now look at the impact of power control error on the capacity of the three multicast schemes. Considering the PTP scheme, Figure 8.13 plots A_v against A_m for four different values of σ_e, the standard deviation of the PCE, $\sigma_e = 0.0\,\text{dB}$, $1.0\,\text{dB}$, $1.5\,\text{dB}$ and $2.0\,\text{dB}$.

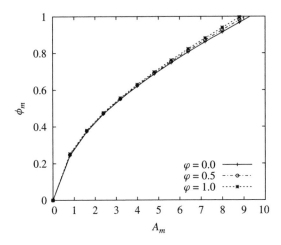

Figure 8.11 ϕ_m against A_m for different values of φ with PTP scheme

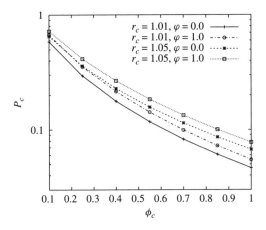

Figure 8.12 P_c against ϕ_c for different values of r_c and φ with PTM scheme

It can be observed that the capacity is gradually reduced by the PCE. The PCE increases the variance of the interference power level, which leads to an increase in the BS outage probability. With $\sigma_e = 2.0$ dB, the Erlang capacity is reduced by as much as 30 % compared with the case with perfect power control and $\sigma_e = 0.0$ dB. This leads us to conclude that the impact of PCE on system capacity is significant.

Similarly for the PTM scheme, Figure 8.14 shows that, while the PTM channel for multicast is not susceptible to PCE owing to the fixed allocation of transmit power, the presence of PCE for voice users degrades the system capacity by a sizeable margin. The capacity difference between PCE with $\sigma_e = 2.0$ and no PCE is in the range of 25 %. This

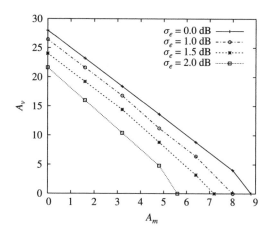

Figure 8.13 A_v against A_m for different values of σ_e with PTP scheme

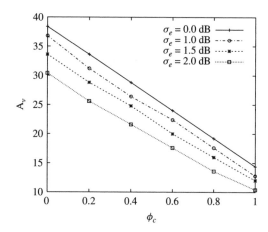

Figure 8.14 A_v against ϕ_c for different values of σ_e with PTM scheme

is consistent with our observation from Figure 8.13. Figure 8.14 is plotted with $\alpha_m = 0.5$ and clearly shows that, even for $\phi_c = 1.0$, a voice capacity of $A_v \approx 12$ Erlang can be supported. This comes by virtue of the statistical multiplexing that is inherent in CDMA systems.

Multicast Bit Energy-to-Interference Requirement

As a final investigation, we now show how the $\gamma_m = E_b/I_0$ requirement of the multicast source affects the capacity of each of the schemes. For the PTP scheme, Figure 8.15 clearly shows that an increase in required γ_m is detrimental to the overall system capacity. The Erlang capacity for multicast with $\gamma_m = 2.0\,\mathrm{dB}$ is nearly 3 times that compared with $\gamma_m = 4.5\,\mathrm{dB}$.

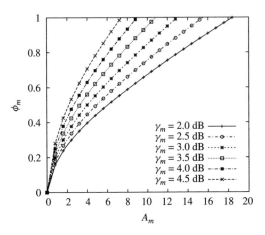

Figure 8.15 ϕ_m against A_m for different values of γ_m with PTP scheme

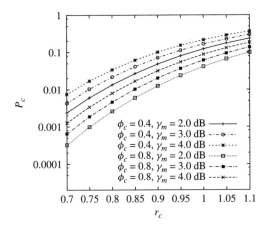

Figure 8.16 P_c against r_c for different values of ϕ_c and γ_m with PTM scheme

Considering the PTM scheme, Figure 8.16 depicts the impact of γ_m on the coverage of the PTM channel. As already observed in Figure 8.15 for the PTP scheme, the impact of an increase in γ_m on the coverage of the PTM channel is significant.

8.6 Summary

In this chapter, we have evaluated the capacity of employing PTP and PTM channels for multicast over the WCDMA air interface in terms of outage probability. Using PTP channels, multicast traffic is transmitted to multicast users on separate dedicated channels, whereas, in using PTM channels, multicast traffic is broadcast to all multicast users within the cell on a single common channel. We have performed a capacity study for both approaches, taking into account deterministic path loss, lognormal shadowing, code orthogonality, power control error and the impact of hard and soft handover. Whereas our capacity evaluation does not consider other important radio impairments such as fast fading, our results provide insight into the performance differences between the two approaches of carrying multicast traffic over the CDMA air interface.

The performance study has shown that the PTP scheme is most suitable when the number of multicast users within a cell is fairly small. With $R_m = 60$ kbps as the physical-layer bit rate for multicast for both the PTP and PTM channels and $\alpha_m = 0.5$ as the activity factor of the multicast source, the PTP scheme in a single-service environment requires the least amount of transmission power for groups consisting of up to $K_m \approx 7$ multicast users within a cell. A fraction of BS transmit power of $\phi_c = 0.8$ allocated to the PTM channel provides suitable coverage, allowing multicast users to perform a hard handover without excessive latency or packet loss, and, purely from a radio perspective, the PTM channel is able to support an arbitrary number of multicast users.

9

Cost Analysis of Multicast Routing[†]

9.1 Introduction

Multicast routing in UMTS and CDMA2000 is managed by having downstream nodes inform their upstream parent nodes that they wish to receive multicast traffic. The registration procedure allows a downstream node to inform its upstream parent node that it wishes to be part of the multicast distribution tree, whereas the deregistration procedure informs an upstream node that a downstream node wishes to leave the multicast distribution tree. The multicast distribution tree needs to be maintained during the session whenever network nodes wish to join or leave the multicast distribution tree.

Alternatively, multicast routing can be achieved by having network nodes store the list of subscribers that belong to a multicast group. The network dynamically forwards multicast packets only to those nodes that are serving multicast subscribers. This may be more efficient in scenarios in which subscribers are highly mobile and re-establishing the entire multicast tree as a result of user mobility is thus costly.

The goal of this chapter is to describe an alternative mechanism for performing multicast routing in UMTS networks and to quantify the performance trade-off between the proposed mechanism for dynamcially routing multicast packets in UMTS networks and the multicast routing mechanism employed in MBMS. We refer to the proposed scheme as dynamic multicast. In addition to evaluating the dynamic multicast mechanism and the routing mechanism employed in MBMS, we also evaluate the cost of two other mechanisms: simple-minded broadcast over the entire network area and multiple unicast where each multicast user receives data using dedicated unicast connections. We quantify the cost of multicast routing by measuring the packet delivery cost and the location update cost. The packet delivery cost measures how efficient the delivery of multicast packets to

[†] Based on "Modeling and Analysis of an Efficient Multicast Mechanism for UMTS", by Rümmler, R.; Yun Won Chung; Aghvami, A. H.; which appeared in IEEE Transactions on Vehicular Technology, Volume 54, Issue 1, January 2005, page(s): 350–365

a group of multicast users is, whereas the location update cost measures how efficiently the mobility of multicast users is handled. The results provide insight into the trade-off between the different mechanisms and show in which scenarios each of the mechanisms performs best.

9.2 Dynamic Multicast for UMTS

With MBMS, multicast routing takes place in a bottom-up fashion. Downstream nodes inform their respective upstream parent nodes that they wish to receive multicast traffic for a particular service. However, this makes it necessary to perform additional procedures for location updates. If location updates due to the mobility or the group dynamics of the multicast members occur frequently, re-establishing the multicast tree can be very costly. In many cases, re-establishing the entire multicast tree is clearly suboptimal.

On the other hand, multicast routing can take place in a top-down fashion, equivalently to how unicast routing is performed, by having each node determine dynamically which downstream nodes wish to receive multicast traffic. In doing so, no changes to the existing location update procedures are required, making the multicast mechanism highly responsive to mobility and group dynamics and also less costly to the network.

Figure 9.1 shows a subset of a UMTS network. In this architecture, there are two SGSNs connected to the GGSN, three RNCs and nine Node Bs. The BM-SC is the origin of multicast traffic. We assume that the BM-SC is colocated with each GGSN. With multicast, only the Node Bs with multicast subscribers should be forwarded multicast

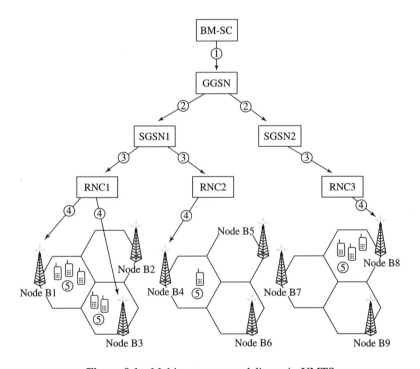

Figure 9.1 Multicast message delivery in UMTS

packets. In Figure 9.1, Node B1, Node B3, Node B4 and Node B8 therefore receive the multicast packets issued by the BM-SC. In order to route multicast packets dynamically within the network, the following steps need to be performed:

1. The BM-SC sends a multicast packet and forwards it to the GGSN that has registered to receive the multicast traffic.
2. The GGSN receives the multicast packet and, by querying its multicast routing records, determines which downstream SGSNs have multicast subscribers residing in their respective service areas. The term *downstream* refers to the topological position of one node with respect to another and relative to the distribution of the multicast data flow. In Figure 9.1, the GGSN duplicates the multicast packet and forwards it to SGSN1 and SGSN2.
3. Both destination SGSNs receive the multicast packet and, having queried their multicast routing records, determine which RNCs are to receive the multicast packets. In doing so, the destination SGSN may page registered multicast subscribers that are PMM-IDLE.
4. The destination RNCs receive the multicast packet and send the multicast packet to the Node Bs that have established the appropriate radio bearers for the multicast application. In Figure 9.1 these are Node B1, Node B3, Node B4 and Node B8. This may involve performing the paging procedure in case the RNC is not tracking any of the multicast UEs on cell level.
5. The multicast subscribers receive the multicast packet over the WCDMA air interface.

It is, however, not trivial to implement the above multicast delivery scheme without modifying the existing UMTS mechanisms for routing and mobility management. More simply, multicast can be based on broadcast by means of flooding or one-to-many unicast with only minor modifications to existing UMTS functionality. With broadcasting, multicast messages are simply flooded to all nodes within the network. Thus, all nine Node Bs in Figure 9.1 receive the multicast packets, even though there are no multicast subscribers in Node B2, Node B5, Node B6, Node B7 and Node B9. With one-to-many unicast, each multicast packet is forwarded separately to each member of the multicast group. In Figure 9.1, Node B1, for instance, would receive three duplicate copies of the same multicast message. Neither broadcast nor one-to-many unicast can therefore be considered to be optimal in terms of bandwidth usage. The efficiency of the one-to-many routing scheme thus depends on the distribution of the group of users within the network.

We now describe our proposed dynamic multicast mechanism that allows for multicast transmission with minimum packet duplication while achieving high responsiveness to mobility and group dynamics. We describe what multicast records are maintained at the respective network nodes, how these records are established and subsequently maintained through the standard location update procedures and how multicast packets are delivered to the subscribers that form the multicast group.

9.2.1 Multicast Tables

In UMTS, a single GTP tunnel is assigned to each user separately. Applied to multicast, this results in each multicast packet being delivered to each multicast user separately. In order to overcome this limitation, two types of record are utilized by the proposed multicast mechanism. At the GGSN and SGSN, we implement a Multicast PDP (M-PDP) context, while at the RNC we implement a Multicast RAB (M-RAB) context for each multicast group being served by the respective node. M-PDP and M-RAB contexts consist of a Multicast Group Context (MGC) and one or more Multicast Subscriber Records (MSRs). The MGC stores parameters that apply to the whole multicast group, such as the multicast group address and QoS profiles. Each MSR, on the other hand, acts as a pointer to the mobility and routing contexts of a single multicast subscriber, with all MSRs accounting for the subset of multicast subscribers that are within the service area of the respective network node. An M-PDP or M-RAB context therefore establishes a mapping between a multicast group address and the subset of multicast group members that are registered at a given GSN or RNC network node. The information elements stored in the MGC consist of the following:

- M-PDP context identifier;
- multicast address (for example 238.255.254.1);
- QoS profile for the multicast group;
- Multicast TEID (M-TEID) for the Gn or Iu interface.

Whereas the M-PDP context identifier uniquely identifies an M-PDP context, the M-TEID uniquely identifies a multicast tunnel over the Gn and Iu interfaces. As with conventional TEIDs, separate M-TEIDs are maintained for the Gn and Iu interfaces.

A requirement of the proposed mechanism is that all multicast subscribers maintain valid PDP and RAB contexts. PDP and RAB contexts establish endpoint addresses for multicast subscribers and are needed in order to route uplink traffic that may result from the multicast application, such as retransmission requests for erroneous packets towards the UMTS gateway. The PDP and RAB contexts maintained for multicast subscribers are updated in the usual fashion according to the standard location update procedures.

The principal operation of the multicast mechanism is performed as follows. Since each MSR functions as a pointer to the actual mobility and routing contexts of multicast subscribers, upon arrival of a multicast packet the GGSN, SGSN and RNC perform a look-up of the actual routing and mobility records associated with each MSR within the M-PDP context. Once the look-up procedure for all multicast subscribers is completed, the node establishes a list of routing addresses for all downlink nodes serving multicast subscribers. The node then forwards a single copy of a multicast packet only to those downlink nodes that are serving multicast subscribers, irrespective of the number of multicast subscribers being served by the node.

In using the multicast subscriber records as pointers to the mobility and routing records of multicast subscribers, the standard location update procedures remain largely unchanged for multicast subscribers. The MSR must allow the GSN and RNC nodes to uniquely identify the mobility or routing contexts belonging to a multicast subscriber. For that purpose, an MSR at the GGSN or SGSN stores the IMSI of the multicast user as well as the Transaction Identifier (TI) of the associated PDP context. While the IMSI uniquely identifies a network subscriber, the TI uniquely identifies a PDP context (3GPP, 2007b). At the RNC, an MSR stores the IMSI as well as the RAB ID, both of which allow the RNC to reference the RAB contexts associated with the multicast application. In order to facilitate location update procedures for multicast subscribers that require the transfer

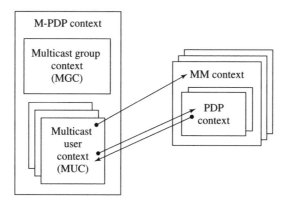

Figure 9.2   Pointers between M-PDP and mobility/routing contexts at the SGSN

of mobility and routing contexts between nodes, each PDP or RAB context associated with a multicast subscriber stores the context identifier of the M-PDP or M-RAB context with which it is linked. The M-PDP or M-RAB context identifiers are inserted into the PDP or RAB routing contexts during group registration and removed when a UE leaves a multicast group.

Figure 9.2 illustrates the pointing mechanism utilized by an M-PDP context maintained at the SGSN. An MSR at the SGSN acts as a pointer to the MM and PDP contexts of a multicast user. Similarly, an M-RAB context at the RNC maintains pointers to RNC and RAB contexts. Since the GGSN does not maintain MM contexts, the MSRs of an M-PDP context at the GGSN only maintain pointers to the PDP contexts of multicast users.

9.2.2   Group Management

We now describe how subscribers register for multicast groups and how the M-PDP and M-RAB context records are established through this procedure. IGMP and MLD are used as group administration protocols in IPv4 and IPv6 respectively. In our scheme, multicast group registration is performed by activating new or modifying existing M-PDP and M-RAB contexts at the GSN and RNC nodes. Subscribers wishing to join a multicast group request the activation of an M-PDP context, equivalently to activating conventional PDP contexts. The successful activation of an M-PDP context requires that subscribers wishing to join a group have a valid PDP context activated prior to group registration.

Activating a M-PDP context mirrors the PDP context activation procedure (3GPP, 2007b). Figure 9.3 illustrates the signalling messages exchanged in the network when a subscriber joins a multicast group. The process of joining a multicast group is summarized as follows:

1. A subscriber with a valid PDP context wishing to join a multicast group sends an M-PDP context activation request to the SGSN, specifying the multicast group address, IMSI, the requested QoS profile and the TI of a valid PDP context belonging to the subscriber.
2. The SGSN validates the request and either creates a new M-PDP context if no M-PDP context with the same multicast address exists or adds an MSR to an existing M-PDP

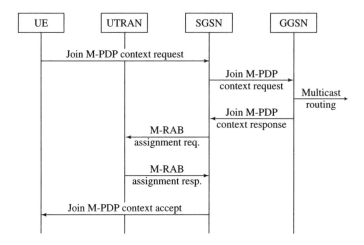

Figure 9.3   Signalling messages for multicast join

context. The SGSN sends an M-PDP context activation request to the GGSN, indicating the M-TEID to be used for the multicast GTP tunnel over the Gn interface.

3. The GGSN either creates a new M-PDP context or adds an MSR to an existing one. If the subscriber is the first to request the activation of the M-PDP context at the GGSN, the GGSN joins the multicast routing tree within external networks based on a suitable multicast routing algorithm. The GGSN confirms the successful activation to the SGSN.

4. The SGSN requests the assignment of an M-RAB context at the RNC. The RNC creates a new M-RAB context if no such context with the same multicast group address exists at the RNC, otherwise an MSR is added to an existing M-RAB context. The RNC confirms the M-RAB assignment procedure to the SGSN, indicating the M-TEID to be used for the GTP tunnel over the Iu interface.

5. The SGSN confirms the successful activation of an M-PDP context to the UE.

Apart from establishing the necessary multicast interworking with external networks in Step 3 above, the signalling messages for joining a group are identical to the signalling messages for activating a PDP context (3GPP, 2007b). Leaving a multicast group is performed by completing the M-PDP context leave procedure, which involves the same message exchanges as the M-PDP activation procedure described above.

9.2.3 *Multicast Mobility Management*

We now describe how the M-PDP and M-RAB contexts are maintained through the standard location update procedures. Owing to the look-up mechanism employed by the proposed scheme in using MSRs as pointers to the actual mobility and routing contexts belonging to multicast subscribers, the standard location update procedures remain largely unchanged.

The service area of an SGSN is partitioned into several RAs. When a UE moves from one RA to another, an RA update is performed, which informs the SGSN of the UE's

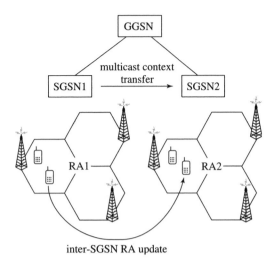

Figure 9.4   Location update for multicast users

current location. The crossing of two RAs within an SGSN area requires an intra-SGSN RA update, while the crossing of two RAs of different SGSNs requires an inter-SGSN RA update. In performing an inter-SGSN RA update, the MM and PDP contexts of a subscriber are transferred between the old and new SGSN nodes.

With the proposed multicast mechanism, location update procedures for multicast users that do not require the transfer of mobility and routing records between nodes, as is the case with intra-SGSN RA updates, remain unchanged.

An inter-SGSN RA update for a multicast user, on the other hand, requires the transfer of the M-PDP context to which a multicast user performing the update belongs between the old and new SGSN nodes. This is illustrated in Figure 9.4. Prior to transferring the mobility and routing contexts to a new SGSN, the old SGSN checks whether the PDP context maintains a reference to an M-PDP context. If this is the case, the old SGSN copies the MGC and the MSR of the multicast subscriber to the new SGSN. An equivalent procedure is carried out when performing the inter-RNS cell or URA update.

9.2.4   Multicast Packet Forwarding

We now describe the processing performed at the GGSN, SGSN and RNC nodes in routing multicast traffic towards the subscribers that form the multicast group. We assume IP multicast as the protocol employed for the multicast application, and that the BM-SC that is colocated with the GGSN forms part of the multicast delivery tree. The steps required to process a multicast message at the GGSN are described as follows:

1. Upon reception of a multicast packet as identified by the multicast IP address, the GGSN searches its table of M-PDP contexts in order to find an M-PDP context with the same multicast address.
2. Having identified the correct M-PDP context, the GGSN iterates through the list of MSRs and performs a look-up of the PDP contexts belonging to the multicast

subscribers. Based on the look-up, the GGSN dynamically creates a list of the SGSN routing addresses that are serving multicast subscribers of the corresponding group.
3. The GGSN forwards exactly one copy of the multicast message to each unique entry in the list of SGSN routing addresses established in Step 2. The multicast packets are tunnelled using the M-TEID for the Gn interface stored in the MGC.

As with multicast IP addresses that share a common prefix, we assume that M-TEIDs can be distinguished from conventional TEIDs by simple inspection. The steps for processing a multicast message at the SGSN are equivalent to the steps performed at the GGSN for processing a multicast message but additionally require the SGSN to perform the paging procedure for multicast users for which the exact location is not known to the SGSN. Multicast packets are subsequently only forwarded to RNCs that are serving multicast subscribers.

1. Having received a multicast N-PDU, the SGSN attempts to correlate the N-PDU with its list of stored M-PDP contexts.
2. Having found a matching M-PDP context, the SGSN iterates through the MSRs encapsulated in the M-PDP context. Each MSR is used to perform a look-up of the actual mobility and routing contexts of the multicast subscriber. If the SGSN does not have valid routing information for the multicast subscriber because it is in PMM-IDLE state, the paging procedure is performed in order to determine the routing address of the actual RNC serving the multicast subscriber. Once the paging procedure has been

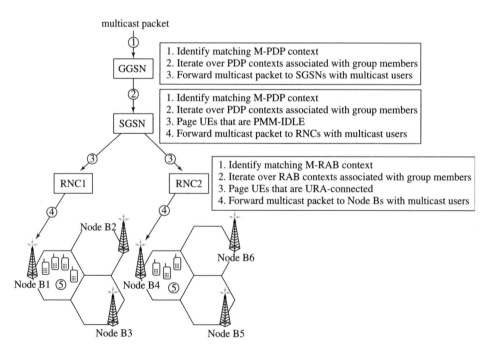

Figure 9.5   Packet forwarding with dynamic multicast

performed for all multicast subscribers that are PMM-IDLE, the SGSN dynamically establishes a list of routing addresses of the RNCs that are serving multicast subscribers.

3. The SGSN forwards exactly one duplicated copy of the multicast packet to each unique entry within the RNC address list created in Step 2. The multicast packet is tunnelled using the M-TEID for the Iu interface.

The RNC is the receiving endpoint of a tunnel from the SGSN. The RNC performs the same look-up procedure as the SGSN in querying the mobility and routing details for all MSRs within the corresponding M-RAB context. Multicast message forwarding within UTRAN across the Iub and Uu interfaces may be performed on PTM or PTP channels, as discussed in Chapter 8.

Figure 9.5 summarizes the major steps performed in routing multicast packets according to the proposed multicast mechanism. In Figure 9.5 we assume that the RNC duplicates and forwards multicast packets only to those Node Bs that have multicast subscribers, and that multicast packets are carried over the air interface on common channels.

9.3   Cost Analysis

In this section, we analyse the cost of multicast in cellular networks. The cost of multicast can be broken down into packet delivery and location update cost. We compare the packet delivery and location update cost of four multicast schemes, namely broadcast, multiple unicast, dynamic multicast and MBMS. Multiple unicast is achieved by duplicating multicast packets at the GGSN for all group members and subsequently forwarding the duplicated packets to each group member separately. We firstly introduce the parameters used for the cost analysis, followed by our models for the multicast user distribution, user mobility and packet traffic. The actual expressions for the packet delivery and location update cost are derived in Appendix B.

9.3.1   Parameters for Cost Analysis

We consider a subset of a UMTS network consisting of a single GGSN and a total of $N_{S/G}$ SGSN network nodes connected to the GGSN. Each SGSN manages a service area consisting of $N_{R/S}$ RAs. Within each RA, a total of $N_{R/R}$ RNC network nodes are located. The service area managed by a single RNC network node in turn is subdivided into $N_{U/R}$ URAs. Finally, each URA consists of $N_{B/U}$ cells. The total number of RAs, RNCs, URAs and cells are then

$$N_{RA} = N_{S/G} \cdot N_{R/S}$$

$$N_{RNC} = N_{S/G} \cdot N_{R/S} \cdot N_{R/R}$$

$$N_{URA} = N_{S/G} \cdot N_{R/S} \cdot N_{R/R} \cdot N_{U/R}$$

$$N_B = N_{S/G} \cdot N_{R/S} \cdot N_{R/R} \cdot N_{U/R} \cdot N_{B/U}.$$

For simplicity, we assume that the BM-SC is colocated with the GGSN.

We distinguish between the cost of signalling messages and packet deliveries. We make a further distinction between processing costs at nodes and transmission costs on links, both for signalling and packet deliveries. Similarly to (Ho and Akyildiz, 1996), we assume that there is a cost associated with each link and node of the network, both for signalling and packet deliveries. Let $D_{(\cdot)}$ denote the transmission cost of packet delivery and $S_{(\cdot)}$ the transmission cost of signalling between two network nodes. Let $p_{(\cdot)}$ denote the processing cost of packet delivery and $a_{(\cdot)}$ the processing cost of signalling at a network node.

Network operators will typically deploy an IP-based backbone network between the GGSN, SGSN and RNC nodes. The links between these nodes will therefore consist of more than one hop. Let l_{hg} be the average distance between the HLR and the GGSN in terms of the number of hops travelled by packets, let l_{gs} be the average distance between the GGSN and the SGSN and let l_{sr} be the average distance between the SGSN and the RNC. The distance between the RNC and the Node B consists of a single hop, therefore $l_{rb} = 1$. Let $h_{(\cdot)}$ denote the unit transmission cost of a given link between two nodes. We weight the transmission costs for packet transmissions by a factor of w_{dt}, and for signalling by a factor of w_{st}. For sake of generality and ease of evaluation, we furthermore weight all transmission costs by a factor of w_t, and all processing costs by a factor of w_p. This allows us readily to evaluate only the transmission or processing costs by selectively setting $w_t = 0$ or $w_p = 0$. By default, the weighting factors w_p and w_t are chosen as $w_p = 1$ and $w_t = 1$ respectively.

The packet transmission cost over a wireline segment of the network can be written as

$$D_{(\cdot)} = w_t w_{dt} l_{(\cdot)} h_{(\cdot)}.$$

The transmission cost for signalling over a wireline segment can be written as

$$S_{(\cdot)} = w_t w_{st} l_{(\cdot)} h_{(\cdot)}.$$

The transmission costs over the air interface for packet transmissions and signalling are $S_{bu} = w_t w_{st} h_{bu}$ and $D_{bu} = w_t w_{dt} h_{bu}$ respectively.

Let $c_{(\cdot)}$ denote the unit processing cost at the HLR, GGSN, SGSN, RNC or Node B respectively. Similarly to the transmission costs for packet deliveries and signalling, we weight the processing costs for packet deliveries and signalling by the factors w_{dp} and w_{sp} respectively. The processing costs at the GGSN, SGSN, RNC and Node B for data traffic are then given by

$$p_{(\cdot)} = w_p w_{dp} c_{(\cdot)}.$$

Similarly, the processing cost at the HLR, GGSN, SGSN, RNC and Node B for signalling are

$$a_{(\cdot)} = w_p w_{sp} c_{(\cdot)}.$$

The cost parameters may be interpreted in a number of ways (Ho and Akyildiz, 1996). For instance, the cost parameters may be expressed in terms of the delay required by a particular network element in processing a packet or a signalling message, or the delay in sending a packet or signalling message through a particular link. On the other hand, the network administrator can assign relative costs to the elements and links of the network on the basis of the current usage and the expenses required to operate a particular network element or link. The intention of this chapter is not to present a method for determining these cost parameters. Instead, we perform the analysis of the proposed multicasting scheme assuming that the above cost parameters are available.

9.3.2 Modelling of Multicast User Distribution

As in (Lin, 2001), we classify the total number of RAs into k_{RA} categories. For $1 \leq i \leq k_{RA}$ there are $N_i^{(RA)}$ RAs of class i, and therefore $N_{RA} = \sum_{i=1}^{k_{RA}} N_i^{(RA)}$. We then define the total number of RAs that have multicast users as (Lin, 2001)

$$n_{RA} = \sum_{i=1}^{k_{RA}} (1 - e^{-\theta_i^{(RA)}}) \cdot N_i^{(RA)}, \tag{9.1}$$

where $\theta_i^{(RA)}$ represents the number of multicast users for the $N_i^{(RA)}$ RAs in class i.

If there are n_{RA} RAs serving multicast users, the probability that an SGSN does not have any such RA is

$$P_{S/G} = \begin{cases} \left. \binom{N_{RA} - N_{R/S}}{n_{RA}} \middle/ \binom{N_{RA}}{n_{RA}} \right. & \text{if } n_{RA} \leq N_{RA} - N_{R/S} \\ 0 & \text{otherwise.} \end{cases} \tag{9.2}$$

Based on (9.2), the average total number of SGSNs that are serving multicast users can be calculated as

$$n_{S/G} = N_{S/G} \cdot (1 - P_{S/G}). \tag{9.3}$$

The total number of multicast users in the network is given by (Lin, 2001)

$$N_m = \sum_{i=0}^{k_{RA}} N_i^{(RA)} \theta_i. \tag{9.4}$$

We assume that all RNCs within the service area of a class i RA have the same multicast population distribution density as the RA. Based on a uniform density distribution within a single RA, the multicast population of an RNC within the service area of a class i RA is

$$\theta_i^{(RNC)} = \theta_i^{(RA)} / N_{R/R}.$$

The total number of RNCs of class i is

$$N_i^{(RNC)} = N_i^{(RA)} \cdot N_{R/R}.$$

With $k_{RNC} = k_{RA}$, the total number of RNCs that have multicast users is given by

$$n_{RNC} = \sum_{i=1}^{k_{RNC}} (1 - e^{-\theta_i^{(RNC)}}) \cdot N_i^{(RNC)}.$$

The same applies to the cells within the service area of an RNC. The average number of multicast users for a single cell of class i is given by

$$\theta_i^{(B)} = \frac{\theta_i^{(RNC)}}{N_{U/R} \cdot N_{B/U}}.$$

The number of Node Bs belonging to class i is

$$N_i^{(B)} = N_i^{(RNC)} \cdot N_{U/R} \cdot N_{B/U}.$$

With $k_B = k_{RNC}$, the total number of Node Bs serving multicast users is

$$n_B = \sum_{i=1}^{k_B} (1 - e^{-\theta_i^{(B)}}) \cdot N_i^{(B)}.$$

9.3.3 Modelling of User Mobility

Mobility management is performed hierarchically by the CN and UTRAN. The SGSN tracks the location of UEs on RA level. Whenever a UE moves to a new RA, it sends an RA update request to its assigned SGSN. An intra-SGSN RA update takes place if the RA the UE has moved to is assigned to the same SGSN as the old RA, whereas an inter-SGSN RA update takes place when the new RA is administered by a different SGSN to the old RA (Bettstetter, Vögel and Eberspächer, 1999). In the latter case, the new SGSN requests the old SGSN to send the MM and PDP contexts of the user. Afterwards, the new SGSN informs the GGSNs involved about the UE's new routing context. The RNC tracks the location of UEs with an active RRC connection on cell or URA level. A cell or URA update that is performed across the boundaries of an RNS is referred to as a Serving RNS (SRNS) relocation. The SRNS relocation procedure involves the transfer of RAB contexts between RNCs as well as a signalling exchange between the new RNC and the SGSNs involved in order to update the routing context at the SGSN.

Assuming that the direction of user movement is uniformly distributed over $[0, 2\pi)$, and using a fluid flow mobility model, the average number of users crossing a boundary of perimeter length l at speed v per unit time is (Mohan and Jain, 1994)

$$M_c = \rho v l / \pi, \tag{9.5}$$

where ρ is the population density per unit area.

We use Equation (9.5) to calculate the average number of crossings per unit time for the areas of an SGSN, RA, RNC, URA and cell. In our model, the population density of multicast users per unit area is defined as

$$\rho_m = \frac{N_m}{A_t} = \frac{N_m}{N_{S/G} \cdot (l_{SGSN}/4)^2},$$

where $A_t = N_{S/G} \cdot (l_{SGSN}/4)^2$ is the total network area.

The average number of crossings for the different areas considered are calculated as

$$M_{SGSN} = \frac{\rho_m \upsilon l_{SGSN}}{\pi} \cdot N_{S/G}$$

$$M_{RA} = \frac{\rho_m \upsilon l_{RA}}{\pi} \cdot N_{RA} - M_{SGSN}$$

$$M_{RNC} = \frac{\rho_m \upsilon l_{RNC}}{\pi} \cdot N_{RNC} - M_{RA} - M_{SGSN} \qquad (9.6)$$

$$M_{URA} = \frac{\rho_m \upsilon l_{URA}}{\pi} \cdot N_{URA} - M_{RNC} - M_{RA} - M_{SGSN}$$

$$M_B = \frac{\rho_m \upsilon l_B}{\pi} \cdot N_B - M_{URA} - M_{RNC} - M_{RA} - M_{SGSN}.$$

9.3.4 Modelling of Packet Traffic

Whenever a PMM-IDLE subscriber receives data, its operational mode is changed to PMM-CONNECTED. If packets in a packet train are close enough, then a subscriber's state will remain PMM-CONNECTED during the packet train transmission. This is referred to as a data session. The same applies to the mobility management performed at the RNC, in that a subscriber will remain cell connected during an active data session. The *holding time* of data sessions can therefore be broadly considered to be analogous to call holding times found in connection-oriented networks (Zhang, Castellanos and Campbell, 2002). Paging therefore only needs to be performed on the first packet of a data session.

9.4 Numerical Results

In this section we briefly discuss the performance of the multicast mechanism in terms of scalability and responsiveness to mobility and group dynamics and then provide numerical results of broadcast, multiple unicast, dynamic multicast and MBMS in terms of packet delivery and location update cost.

In order to minimize multicast routing state and thus improve the scalability of IP multicast routing protocols, multicast routers only know whether or not there are any receivers for a multicast group on an interface, and not the number or the identity of the receivers (Lee and Miller, 1999). Multicast routers are therefore only aware of which of their immediate neighbours participate in the multicast group. In comparing multicast routing approaches, shared-tree multicast routing protocols generally require less multicast routing state than shortest-path multicast trees and therefore scale well over wide-area networks (Deering *et al.*, 1996).

In mobile networks, however, owing to the requirements of managing mobility as well as subscriber administration (for example, authorization, authentication and accounting), each UMTS node maintains per-subscriber records that are created as the subscriber requests the establishment of a virtual connection across the network. The multicast tables only maintain pointers to these subscriber records and therefore by themselves require very little additional state, therefore increasing the scalability of the mechanism.

A multicast routing protocol should not only allow multicast members to join and leave the multicast tree in a seamless fashion but also ensure that a join or leave event does not require widespread changes to the routing tables in the network (Sahasrabuddhe and Mukherjee, 2000). In terms of responsiveness to group dynamics, group-shared trees typically require that the multicast tree is re-established if group dynamics change, which is usually not the case for shortest-path multicast trees.

Since the proposed multicast mechanism dynamically adapts the multicast tree to changes in group membership, group dynamics are only subject to signalling delays in updating the multicast tables within the network nodes and do not require costly changes to the multicast tree structure. Similarly, the movement of multicast users does not require static changes to the multicast tree since multicast packets are routed to the appropriate downlink nodes upon querying the unicast routing records, which are updated according to location changes in the usual fashion.

In considering stationary hosts over the Internet, the advantages of multicast over broadcast and multiple unicast are usually evaluated in terms of bandwidth consumption and processing requirements at the end-hosts. In mobile networks such as UMTS, however, the differences between the various approaches must take into account the specific nature of mobility management in UMTS. The use of broadcast within appropriately defined service areas consisting of one or more cells, in which a network operator offers a particular service such as video or other multimedia, is a viable technique since, firstly, location tracking of multicast users (including paging of multicast users that are idle) is no longer necessary within the service area, and, secondly, packet forwarding at intermediate nodes may be achieved by flooding packets to downstream nodes. Multiple unicast, on the other hand, requires no changes to existing protocols, may provide optimized usage of resources over the air interface and, assuming that bandwidth within the UMTS core network is plentiful, may provide satisfactory performance in most cases.

For the performance evaluation, we assume that the cost of transmitting paging messages or packets and the processing costs at the respective nodes are available. Furthermore, we assume that $l_{(\cdot)}$, the number of hops packets travel, are fixed numbers (Xie and Akyildiz, 2002). The TTL field in the IP header is usually initialized by the sender to 32 or 64 (Stevens, 1994), and therefore the upper limit on the number of hops through which a packet can pass is 32 or 64. With this in mind, we set $l_{hg} = 5$, $l_{gs} = 15$ and $l_{sr} = 10$. The link between an RNC and a Node B consists of a single hop, and therefore $l_{rb} = 1$.

Table 9.1 lists the unit processing and transmission costs as well as the weighting factors used in our performance evaluation. The unit processing and transmission costs in the CN are generally lower than the equivalent costs in UTRAN. Furthermore, we take the transmission cost over the wireless link to be significantly higher than the unit distance transmission cost over a wireline link.

Table 9.1 Parameters for performance analysis

Unit transmission costs					Unit processing costs					Weighting factors					
h_{hg}	h_{gs}	h_{sr}	h_{rb}	h_{bu}	c_h	c_g	c_s	c_r	c_b	w_{dt}	w_{st}	w_{dp}	w_{sp}	w_t	w_p
0.2	0.2	0.2	0.5	5	0.5	0.5	0.5	0.75	1	0.6	0.2	0.1	0.1	1	1

As in Lin (2001), we consider two classes of RAs. A class $i = 1$ RA has a multicast user population of $\theta_1 = 1/\delta$, and a class $i = 2$ RA has a multicast user population of $\theta_2 = \delta$. If $\delta \gg 1$, the class $i = 1$ RA has a small multicast user population and the class $i = 2$ RA has a large multicast user population. Let α be the proportion of the class $i = 1$ RAs, and $1 - \alpha$ the proportion of the $i = 2$ class RAs (Lin, 2001). Each RA of class $i \in \{1, 2\}$ in turn is subdivided into $N_{R/R}$ RNCs of the same class i. Similarly, each RNC of class $i \in \{1, 2\}$ is subdivided into $N_{U/R} \cdot N_{B/U}$ Node Bs of the same class i. For the network topology, we consider a network with $N_{S/G} = 10$, $N_{R/S} = 10$, $N_{R/R} = 10$, $N_{U/R} = 5$ and $N_{B/U} = 5$. We denote the arrival rate of data sessions as λ_s, and the number of packets in a data session as N_p.

9.4.1 Packet Delivery Cost

We now evaluate the packet delivery cost of broadcast, multiple unicast and dynamic multicast. This is followed by a performance study of MBMS compared with dynamic multicast. The expressions for the packet delivery cost are derived in Appendix B.

Figure 9.6 plots the cost of broadcast, multiple unicast and dynamic multicast against α. We set $\delta = 100$ and $N_p = 5000$, which corresponds to a multicast session consisting of a large number of packets. We observe from Figure 9.6 that the cost for broadcast is constant, whereas the costs of the other schemes decrease as α increases. With $\delta \gg 1$, it is likely that there are no multicast members in an RA of class $i = 1$, and that there are many members in an RA of class $i = 2$ (Lin, 2001). As α increases, the number of class $i = 1$ RAs with a small multicast population increases, whereas the number of class $i = 2$ RAs with a large multicast population decreases. As a result of the increasing density of multicast users within a small number of RAs, the costs for multiple unicast and dynamic multicast decrease as α increases. In Figure 9.6, dynamic multicast has the lowest cost compared with the other two schemes, with the cost of the three schemes converging as α approaches unity.

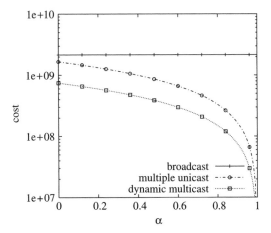

Figure 9.6 Cost of packet delivery against α for dynamic multicast

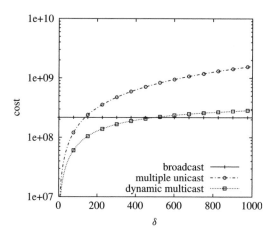

Figure 9.7 Cost of packet delivery against δ for dynamic multicast

Figure 9.7 plots the cost against δ, with $\alpha = 0.1$ and $N_p = 500$. Our first observation from Figure 9.7 is that the cost curve for broadcast is constant, whereas the cost curves for multiple unicast and dynamic multicast are increasing functions of δ. From our earlier discussion, a class $i = 1$ RA has $\theta_1 = 1/\delta$ multicast users, whereas a class $i = 2$ RA has $\theta_2 = \delta$ multicast users. As δ increases, the user population within the class $i = 2$ RAs increases, thereby resulting in an increase in the associated cost for multicast packet delivery. From Figure 9.7 we can see that dynamic multicast has the lowest cost for $\delta < 400$, whereas broadcast has the lowest cost for $\delta > 400$. We furthermore observe that multicast unicast performs well for $\delta < 200$, but is outperformed by broadcast for $\delta > 200$.

Figure 9.8 shows the cost of the three schemes against the number of packets N_p in a single multicast session for $\alpha = 0.1$ and $\delta = 100$. As expected, the cost of the three

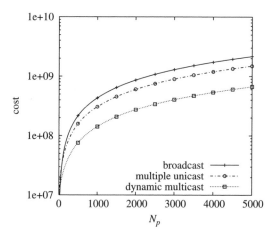

Figure 9.8 Cost of packet delivery against N_p for dynamic multicast

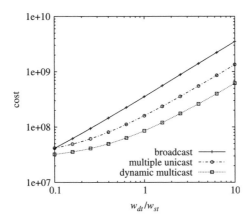

Figure 9.9 Cost of packet delivery against w_{dt}/w_{st} for dynamic multicast

schemes increase as N_p increases. Dynamic multicast has the lowest cost over the entire range of N_p, followed by multiple unicast and broadcast.

Figure 9.9 plots the packet delivery cost against w_{dt}/w_{st} in the range [0.1 : 10]. The ratio w_{dt}/w_{st} relates the cost for packet transmissions against the cost for signalling. For $w_{dt}/w_{st} > 1$, packet transmissions are more costly, whereas the opposite is the case for $w_{dt}/w_{st} < 1$. Figure 9.9 shows that only the cost of broadcast progresses in a linear fashion on a logarithmic scale. Dynamic multicast has the lowest cost for the entire range [0.1 : 10], with broadcast having the highest cost.

We now compare the packet delivery cost of MBMS against that of dynamic multicast. Figure 9.10 plots the cost curves of multiple unicast, our multicast scheme and MBMS against α, with $N_p = 500$ and $\delta = 100$. The curve for multiple unicast is included for reference purposes. Figure 9.10 shows that MBMS has a slightly lower cost than dynamic multicast over the entire range of α, whereas multiple unicast has the highest cost of the three schemes.

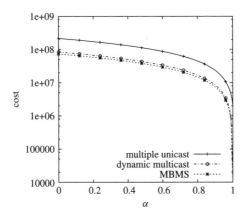

Figure 9.10 Cost of packet delivery against α for MBMS

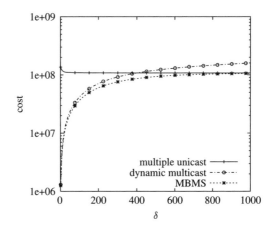

Figure 9.11   Cost of multicast packet delivery against δ for MBMS

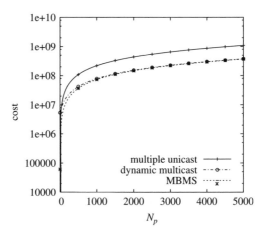

Figure 9.12   Cost of multicast packet delivery against N_p for MBMS

Figure 9.11 plots the cost curves for the three schemes against δ. The cost of MBMS and dynamic multicast are very similar for $\delta < 100$. For $\delta > 100$, MBMS has a lower cost than our scheme. Interestingly, for $\delta > 400$, multiple unicast outperforms dynamic multicast. Similarly, the cost of MBMS approaches that of multiple unicast for $\delta \approx 1000$.

Figure 9.12 plots the cost curves for the three schemes against N_p. The difference between the cost of MBMS and dynamic multicast is negligible over the entire range of N_p.

Figure 9.13 depicts the cost curve against w_{dt}/w_{st}. The curves for all three schemes increase linearly on a logarithmic scale, with the gap between the three curves remaining constant throughout. This is not the case in Figure 9.13, where the cost of dynamic multicast is the highest for $w_{dt}/w_{st} > 0.2$ but then approaches the cost for MBMS for $w_{dt}/w_{st} > 3$. This behaviour was also observed in Figure 9.9 and is a result of the decreasing impact of paging on the cost of dynamic multicast for $w_{dt}/w_{st} > 1$.

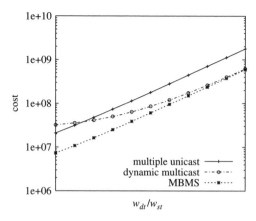

Figure 9.13   Cost of packet delivery against w_{dt}/w_{st} for MBMS

9.4.2   Location Update Cost

In this section, we investigate the location update cost per unit time of MBMS against that of the mobility procedures that are used for dynamic multicast. The expressions for the location update cost are derived in Appendix B.

Figure 9.14 shows the location update cost per unit time for a single multicast user against v. Both curves increase with v, as more area crossings occur as an increasing function of v. Figure 9.14 clearly shows that the location update cost per unit time of dynamic multicast is less than the location update cost per unit time of MBMS. This allows us to conclude that the location update cost of dynamic multicast is generally lower than that of MBMS.

Figure 9.15 plots the location update cost per unit time against the cell perimeter, l_B. As l_B increases, the location update cost per unit time decreases. This is intuitive, since a small cell perimeter translates into frequent cell crossings and thus into a large location

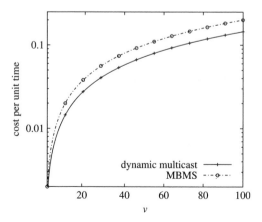

Figure 9.14   Location update cost per unit time against v

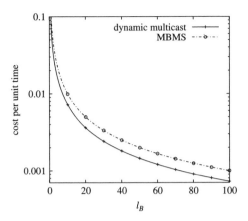

Figure 9.15 Location update cost per unit time against l_B

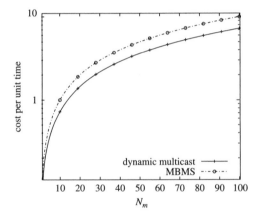

Figure 9.16 Location update cost per unit time against N_m

update cost per unit time. As with Figure 9.14, the location update cost per unit time of dynamic multicast is lower than that of MBMS.

Our final result for the location update cost is shown in Figure 9.16, which plots the location update cost per unit time against N_m. Figure 9.16 is plotted under the assumption that all group members are sparsely distributed over the service area, so that downstream nodes always have to perform the MBMS registration and de-registration procedures towards their respective upstream nodes. Figure 9.16 shows that the location update cost per unit time is an increasing function of N_m, and that, as with the previous two graphs, dynamic multicast has a lower cost than MBMS.

9.5 Summary

In this chapter, we have presented a dynamic multicast mechanism for UMTS that achieves multi-destination packet delivery by establishing multicast tunnels between network nodes along the paths to all receivers. Multicast packets are routed on a per-packet basis to downstream nodes serving multicast users. The implementation and execution of the multicast tables maintained at the nodes are very efficient and require little state. Furthermore, the mechanism can be implemented with only minimal modifications to the standard location update procedures.

In contrast to MBMS, which establishes a multicast routing tree that requires modification as a result of mobility, the proposed scheme responds well to mobility and group dynamics since packet forwarding is performed on a per-packet basis. Our numerical results have highlighted the trade-off between MBMS and the proposed multicast scheme. We have shown that, whereas MBMS does not perform paging and consequently has a slightly lower packet delivery cost, the proposed scheme has a lower location update cost than MBMS since the multicast scheme relies on paging in order to locate multicast users. Also, our multicast scheme does not have to re-establish the entire multicast tree as a result of mobility or group dynamics. Therefore, our scheme performs favourably in terms of location update cost to the network.

10

Parity-Based Reliable Multicast

10.1 Introduction

In this chapter, we investigate the performance of different reliability mechansims for multicast in third-generation networks such as UMTS. We consider reliability mechanisms that combine packet-based FEC with ARQ. We analyse the performance of these mechanisms, applied both to the RLC layer and to the application layer, in terms of channel efficiency, residual packet error rate and delay. As a first step, the residual packet error rates of the RLC layer are established, assuming that the RLC layer is not fully reliable. For multicast over the CDMA air interface, we distinguish between the use of PTM channels that are broadcast over the entire cell and separate PTP channels that are transmitted to individual users of the group within the cell. Subsequently, the results of the RLC layer are used as inputs to determine the performance of the end-to-end reliable multicast mechanism operating on the application layer between the sender and all receivers. Our study provides insight into the cross-layer dependencies between the reliability mechanisms operating on the RLC and the application layer.

10.2 Loss Recovery for Reliable Multicast

For many types of multicast application, such as one-to-many file distribution and software download, reliable multicast delivery is a must. A common way to achieve reliability is to use ARQ. Using ARQ, the sender retransmits lost data upon request from the receiver. The receiver uses a back-channel to send requests for retransmission of lost packets. With ARQ, the transmitter sends packets consisting of payload and error detection codes. At the receiver side, based on the outcome of the error detection procedure, acknowledgement messages are sent back to the transmitter (ACK or NACK, according to the result of error detection). The sender performs packet retransmissions based on such acknowledgements (Rossi, Badia and Zorzi, 2003a). With multicast, standard ARQ mechanisms achieve poor throughput when the number of receivers is large. Also, receiving feedback on the back-channel for a large number of receivers can cause problems at the sender.

Multicast in Third-Generation Mobile Networks   Robert Rümmler, Alexander Gluhak and A. Hamid Aghvami
© 2009 John Wiley & Sons, Ltd

As an alternative to using a back-channel for reliable multicast, a data carousel approach can be used (Acharya, Franklin and Zdonik, 1995). With this approach, the sender continually cycles through and sends out data packets, with the receivers continually receiving packets until they have a copy of each packet. This approach does not require a back-channel but may be wasteful of bandwidth.

FEC can be used as a reliability mechanism to augment or replace other reliability mechanisms. In the general literature, FEC refers to the ability to overcome both erasures (losses) and bit-level corruption (Luby *et al.*, 2002c). With reliable multicast, the primary application of FEC codes is as an erasure code.

The basic principle of FEC-based protocols is that the original data is encoded to obtain some parity data. The original data along with additional parity data can be used to reconstruct the original data if some of it is lost (Nonnenmacher, 1998). The parity data is either sent along with the original data or upon request from the receivers in order to repair different losses at the receivers. The error correction is *forward* in the sense that no feedback from the receiver to the sender or further transmission by the sender is required.

10.2.1 Mechanisms for Parity-Based Loss Recovery

With multicast, no amount of FEC can always ensure reliable data delivery to a large group of receivers. Many reliable multicast protocols rely on combining different techniques such as ARQ and FEC to achieve efficiency and scalability. Combining ARQ and FEC can be done in several ways. With H-ARQ type 1, the transmitter simply adds parity packets generated by an erasure code to a group of data packets in order to reduce the packet error rate, with erroneous data packets being retransmitted in the usual way as with conventional ARQ. Proactively adding parity packets results in a lower probability of packet loss compared with the case without parity packets. Retransmitting the original packets in the case of packet losses has the advantage that the protocol layers responsible for FEC and for reliable multicast remain separated and that packet delays are kept at a minimum. This scheme does not require a large buffer at the receiver but is not optimal in terms of channel efficiency.

With H-ARQ type 2, a further improvement is achieved by transmitting additional parity packets as a result of packet losses instead of retransmitting erroneous data packets, since parity packets are able to repair different packet losses at the receivers. However, this necessitates the integration of FEC and ARQ into a single layer, which in turn increases complexity and also increases the packet delay. The receiver stores the received erroneous packets in a buffer in order to reuse them in combination with the packets from subsequent retransmissions.

A data carousel approach can be taken if no back-channel is available. With the data carousel, the sender continually cycles through and sends out source packets (data packets only or data followed by parity packets), with the receivers continually receiving packets until they have a copy of each packet.

We investigate several schemes that combine reliability strategies such as H-ARQ and the data carousel approach. We assume that, for schemes that rely on a back-channel, feedback can be managed and does not result in feedback implosion at the sender.

We consider erasure codes such as Raptor or Reed-Solomon codes (McAuley, 1990; Rizzo, 1997) that operate by taking k blocks or data packets and producing h parity packets. The k data packets are referred to as a Transmission Group (TG), whereas the combined total of $n = k + h$ data and parity packets are referred to as an FEC block (Nonnenmacher, 1998). At the receiver side, it is sufficient if k of the n packets in the TG are received correctly in any order to reconstruct the original k packets.

HARQ1

With the first scheme we consider, the sender proactively transmits parity in the initial transmission, and retransmits original data packets if packets are received in error. This scheme has very low implementation complexity but does not maximize channel efficiency. We refer to this scheme as HARQ1. In more detail, HARQ1 operates as follows:

1. The packet-based encoder takes k data packets and adds $a = h$ parity packets to the TG.
2. The FEC block consisting of $n = k + a$ packets is transmitted over the channel.
3. If the receiver receives k data packets, then no decoding is necessary. If any k of the data and parity packets are received, then the entire TG can be decoded.
4. If the receiver detects fewer than k correct packets in a given FEC block, the receiver sends a NACK back to the sender, including the TG identifier as well as the identifiers of the erroneous packets.
5. The sender includes the original data packets for which retransmissions are requested in a new TG and transmits the TG, along with a parity packets, to all receivers.

HARQ2

Transmitting additional parity packets in response to packet errors improves the bandwidth requirements of reliable multicast considerably. With HARQ2, the sender transmits only a single additional parity packet as a result of packet errors in a TG. In more detail, HARQ2 operates as follows:

1. The sender transmits k data packets together with a parity packets to all receivers.
2. If the receiver receives fewer than k data packets, the receiver sends a NACK back to the sender, specifying the TG identifier as well as the identifiers of the erroneous packets.
3. The sender transmits exactly one parity packet for each TG for which it has received a NACK.
4. Each receiver attempts to decode the FEC block using the additional parity packets transmitted from the sender.
5. If no further parity packets are available at the sender because all parity packets in an FEC block have been transmitted, the sender starts to retransmit original data packets, as with HARQ1.

HARQ3

With HARQ3, the sender always transmits the maximum required amount of parity packets over the whole receiver population in order to reduce delays as much as possible. All receivers perform the same procedure as with HARQ2. The sender, on the other hand, transmits the maximum number of parity packets that are required to repair packet losses at the receivers. Consider that each receiver i has experienced r_i packet losses in a single TG. The sender collects incoming NACKs, calculates r_i for each receiver, based on the identifiers of the lost packets within each NACK, and sends exactly $R_{max} = \max(r_i)$ additional parity packets to the receivers (Rossi, Fitzek and Zorzi, 2003). This procedure is repeated until all receivers are able correctly to decode the k packets, or until all h parity packets are used up at the sender, which then necessitates the retransmission of original data packets.

HARQ4

Finally, with the fourth scheme we consider, there is no back-channel and therefore the sender transmits parity packets immediately following the original data packets. Receivers then leave the multicast group as soon as they have received all required data packets. This data carousel mechanism, which we refer to as HARQ4, avoids any feedback implosion problems at the sender. When a receiver has received enough parity packets for the TG, it leaves the multicast group and stops receiving packets. This mechanism does not rely on feedback, but requires the receivers to be able to leave the multicast group quickly in order to avoid redundant transmissions of parity packets.

As a final comment, since proactively adding parity packets with HARQ2 and HARQ3 leads to worse performance when the number of receivers is small (Nonnenmacher, 1998), we generally do not consider the proactive sending of parities and $a = 0$.

10.2.2   Reliable Multicast for MBMS

With MBMS, multicast packets originate at the BM-SC and then travel between the GGSN and SGSN to the RNC. The RNC transmits to the base station or Node B, which finally carries the transmission over the air interface, either on PTP or PTM channels.

Reliability mechanisms for multicast are typically implemented on the link layer or on the application layer on an end-to-end basis between the multicast source and all receivers. In MBMS, FEC protection is provided at the application layer between the source and all receivers on the level of individual IP packets. The RLC layer operates in unacknowledged or unreliable mode, meaning that no retransmissions are performed for packets that are lost over the air interface. Residual packet errors of the RLC layer must therefore be handled by the reliability mechanism operating on the application layer.

Raptor coding is implemented in software at the transmitter and at all receivers. Raptor codes are FEC codes that operate to correct erasures (losses), protecting data by encoding the original source data at the sender such that redundant data is added, with the redundant data being able to repair different losses at the receivers (Shokrollahi, 2006). Raptor coding on the application layer responds both to packet losses on the wired part of the cellular network and to residual packet errors of the RLC layer. In MBMS, packet-level protection on the application layer is carried out without a back-channel, and guaranteed reliability can therefore not be provided. An extra file repair procedure used after the actual MBMS transmission ensures error-free file delivery (3GPP, 2007e).

With MBMS, Raptor coding used on the application layer complements the physical-layer FEC protection, recovering those IP packets that have been lost because the receiver was unable fully to correct the received data (3GPP, 2007e). Raptor codes are one of the first known classes of fountain codes with linear time encoding and decoding. Fountain codes (also known as rateless erasure codes) are a class of erasure codes with the property that a potentially limitless sequence of encoding symbols can be generated from a given set of source symbols such that the original source symbols can be recovered from any subset of the encoding symbols of size equal to or only slightly larger than the number of source symbols. It should be noted that a *symbol* can be relatively large, from several bytes to several hundred bytes. Often *symbol* and *packet* are used interchangeably, that is, one symbol is transmitted in each packet.

10.3   Performance Evaluation Method

We investigate the performance of multicast reliability mechanisms operating on the link as well as on the application layer by means of simulation. The simulation tool was custom-developed in a high-level programming language. For each simulation run, $I \geq 10^5$ packets are transmitted from the source to all receivers. At the end of each simulation run, all performance results are calculated and output to file for post-processing within a data plotting application.

We investigate the performance of reliable multicast both on the RLC layer and on the application layer between the sender and all receivers. The RLC transmission/retransmission process at the sender operates by segmenting higher-layer packets or SDUs into smaller PDUs. The PDUs are processed by adding a header and a CRC field, which is needed at the receiver to check the correctness of each PDU. At the receiver, the link-layer PDUs are used to reconstruct higher-layer SDUs. The application layer operates in a similar fashion by breaking application data into packets and adding both a header and a CRC field for correctness verification.

10.3.1   Performance Metrics

For the RLC layer, we consider three performance metrics: multicast throughput, residual packet error rate and packet delay. For reliability on the application layer, we consider both the mean number of packets transmitted (including parity packets) for each received packet and the packet delay as our performance metrics.

Channel Efficiency

Channel efficiency can be expressed by the mean number of transmitted packets (including parity packets) for each received packet as well as by the multicast throughput. We use both expressions for measuring the channel efficiency.

Let M' denote the number of times an arbitrary data packet is transmitted before all multicast group members have received it. We use simulations to obtain $E[M']$. In order to measure the efficiency of parity-based error recovery techniques, the transmission of parity packets needs to be accounted for. Let $E[M]$ denote the mean number of packets transmitted, including parity packets, before all group members have received an arbitrary data packet. Using $E[M']$, which is obtained through simulations, $E[M]$ may be calculated as

$$E[M] = \frac{k + E[H]}{k} \cdot E[M'], \tag{10.1}$$

where $E[H]$ is the average number of parity packets transmitted per TG. In the case of HARQ1, $E[H] = h$, whereas with HARQ2, HARQ3 and HARQ4 the number of additionally transmitted parity packets is obtained by simulation. Considering transmissions without redundancy, the mean number of transmissions is simply $E[M] = E[M']$. The channel efficiency can be measured in terms of the mean number of packets transmitted (including parity packets) before all group members have received an arbitrary data packet.

Let ε_{eff} denote the effective packet error rate. It may be written as

$$\varepsilon_{eff} = 1 - \frac{1}{E[M]}. \tag{10.2}$$

Multicast throughput may be calculated as

$$\tau = 1 - \varepsilon_{eff} = \frac{1}{E[M]}.$$

Residual Packet Error Rate

Assuming that the number of allowed retransmissions for each link-layer PDU is finite, the RLC layer will not be fully reliable. The residual PDU error rate as a function of the maximum number of allowed retransmissions for each PDU is therefore a relevant performance metric. With respect to the end-to-end performance of the reliable multicast mechanism, the residual error rate of an aggregate of K PDUs that constitute a single SDU is important, since any residual SDU or packet errors will trigger retransmissions from error recovery mechanisms operating above the RLC layer. An SDU is considered to be erroneous if at least one PDU constituting the SDU is erroneous.

Packet Delay

An important metric of the RLC layer is the delay of a single PDU. As described in (Chang and Yang, 1994), (Kim and Krunz, 2000) and (Rossi, Badia and Zorzi, 2003b), the overall PDU delay with an ARQ protocol can be subdivided into three contributions. The first contribution is due to the queueing delay in the source buffer, in other words the time between the PDU release by higher layers and the instant of its first transmission over the channel. The second contribution is the time between the first transmission and the correct reception of the PDU, which depends only on the channel behaviour. The final delay contribution is the time spent in the receiver resequencing buffer. Even though the sender transmits packets in order, they can arrive out of sequence as a result of random errors and subsequent retransmissions. Correctly received PDUs must wait in the resequencing buffer until all PDUs with a lower identifier have been correctly received.

As is usually done in the literature, the three delay terms are referred to as *queueing*, *transmission* and *resequencing* delay. In the following, the delivery delay, defined as the time between the first transmission of the packet and its successful release from the resequencing buffer, will be investigated. This definition of delivery delay corresponds to the sum of the transmission and resequencing delays. Our evaluation of the delivery delay relies on the assumption that, once a packet is correctly transmitted, a new packet is always available for transmission. Should this assumption not hold, the delivery delay results correspond to an upper bound or worst case (Rossi, Badia and Zorzi, 2003a).

In our model, the time is slotted and the slot time corresponds to a single PDU transmission. The value of the round-trip time corresponds to an integer number of slots m. An SDU is considered as a higher-layer packet that is composed of K link-layer PDUs (Rossi, Fitzek and Zorzi, 2003). For the purposes of our study, an SDU corresponds to an IP packet.

From the viewpoint of higher layers, the delivery delay of an aggregate of K PDUs constituting a single SDU is important. Here, the SDU delivery delay is the number of slots elapsed between the slot where the first PDU of a given RLC SDU is transmitted for the first time over the channel and the slot where the receiver correctly receives the last PDU composing the SDU (Rossi, Badia and Zorzi, 2003b). As described in (3GPP, 2008d), SDUs may be delivered in sequence or out of sequence. We consider in-sequence delivery of SDUs to upper layers.

10.3.2 Simulation Approach

Simulation results are obtained in two stages: firstly, all relevant performance metrics of the RLC layer are calculated, and then these results are used as inputs in order to obtain the performance metrics of the reliability mechanism on the application layer.

Our simulation approach is illustrated in Figure 10.1. As a first step, we use nominal values for the rate and average burst length of the PDU error process as input parameters for evaluating the statistics of the corresponding SDU residual packet error process. By means of simulation, we obtain the rate and average burst length of the SDU error process. The resulting values for the rate and average burst length of the SDU error process obtained in the first simulation run are then used as input parameters for the second simulation run.

In the second simulation run, we obtain the mean number of transmissions per received packet, as well as the packet delay of the reliability protocol on the application layer.

We model the interactions between the RLC and application layer by operating the application layer at the endpoints of the RLC layer. This is justified, as we are only interested in the performance metrics of the application layer in correcting residual packet errors of the RLC layer. Figure 10.2 illustrates our modelling approach. As can be seen from Figure 10.2, the BM-SC and the RNC are colocated such that the application layer only sees residual packet errors of the RLC layer. As a direct result of this approach, we only model a single multicast group, with both the RLC and application layer servicing the same multicast users.

Our approach for evaluating the performance of the application layer with PTP or PTM channels is as follows. From the viewpoint of the application layer, employing a single PTM channel with parity-based loss recovery for a multicast group is equivalent to having a single receiver with a residual packet error rate that is dependent on the number

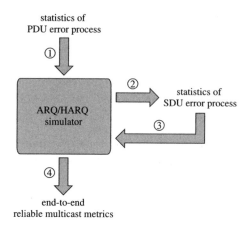

Figure 10.1   Simulation methodology for evaluating end-to-end reliable multicast

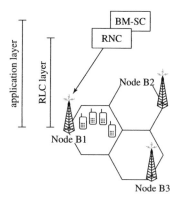

Figure 10.2   Model for analysing interactions between RLC and application layer

of receivers using the PTM channel. With multiple PTP channels without FEC, on the other hand, the residual packet error rate remains constant for each receiver, with each multicast receiver contributing independently to packet retransmissions of the end-to-end reliable multicast protocol.

10.3.3   Packet Error Model

Packet transmissions are characterized by means of a two-state Discrete-Time Markov Chain (DTMC), with states 0 and 1. We assume that transmissions in state 1 are always erroneous, whereas state 0 is error free (Rossi, Badia and Zorzi, 2005). This model captures the memory that is inherent in bursty channels such as wireless fading channels. The error process transition probability matrix P is

$$P = \begin{pmatrix} p_{00} & p_{01} \\ p_{10} & p_{11} \end{pmatrix},$$

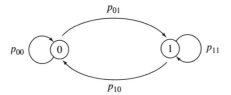

Figure 10.3   Model of a two-state Discrete-Time Markov Chain (DTMC)

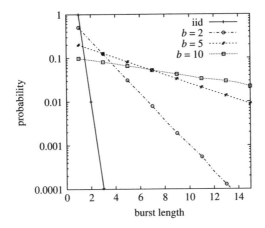

Figure 10.4   Burst length distribution at one receiver for $p = 0.01$

with the corresponding i-step transition probability matrix $\boldsymbol{P}(i)$ defined as

$$\boldsymbol{P}(i) = \boldsymbol{P}^i = \begin{pmatrix} p_{00}(i) & p_{01}(i) \\ p_{10}(i) & p_{11}(i) \end{pmatrix}.$$

The steady-state channel error probability is $p = p_{01}/(p_{10} + p_{01})$, and the average burst length is given by $b = 1/p_{10}$. When errors are distributed in an independent and identically distributed (iid) manner, $p = p_{11} = p_{01}$, so that $b = 1/(1 - p)$.

Figure 10.3 depicts the model of the two-state DTMC. Figure 10.4 illustrates the burst length distribution for different values of b. Figure 10.4 has been plotted by means of simulation. It can be seen that both distributions decrease linearly on a logarithmic scale.

10.4   Reliable Multicast over the Air Interface

MBMS can be delivered using PTP or PTM channels. Switching between both channels depends on the number of users in a cell. With small user populations, PTP or unicast channels are more efficient than PTM channels. With MBMS, switching between PTM and PTP channels can be done flexibly, depending on the number of users in a cell. If a small number of users in a cell request an MBMS service, using multiple PTP or unicast channels is more efficient than using a single PTM channel. This decision is made by counting the number of users requesting the multicast service in a cell–the procedure for doing this is referred to as *counting* (3GPP, 2008a).

In MBMS, the RLC layer is operated in unacknowledged or unreliable mode. This is inefficient if the number of errors is large, as these errors must then be handled by the reliability mechanism operating on the application layer. We investigate the performance of operating the RLC layer in acknowledged or reliable mode for multicast over the air interface. For PTP channels, regular ARQ can be used to provide reliability, whereas, with the PTM channel, parity-based reliability mechanisms are required in order to achieve scalability. Using separate PTP channels, therefore, does not require any changes to the existing ARQ mechanisms, whereas, with PTM channels, parity-based loss recovery mechanisms to the RLC layer must be employed when reliability should be provided at the RLC layer.

10.4.1 PTP Channels for Multicast

We initially consider a transmitter and a single receiver that communicate data packets through a wireless link on a PTP channel with a limited number of retransmission attempts to counteract channel impairments. The sender continuously transmits new PDUs in increasing numerical order as long as ACKs are received. After each PDU reception, the receiver checks for PDU errors and replies with an ACK/NACK accordingly. When a generic PDU is transmitted, say PDU i, the sender must await an ACK message for that PDU until after it finishes the transmission of up to $m - 1$ subsequent PDUs, which can be new PDUs or retransmissions of erroneous PDUs. Unless noted otherwise, we assume a nominal value of $m = 10$ (Rossi, Badia and Zorzi, 2003a). We use a value K = 14 for the SDU or IP packet segmentation (Rossi, Badia and Zorzi, 2003b). Assuming a minimum IP packet size of 576 bytes, the average PDU size is 41 bytes.

Residual Packet Error Rate

Let L be the maximum number of allowed retransmissions for each link-layer PDU. With L finite, not all PDU errors can be handled, thus rendering the RLC layer not fully reliable. Let ε denote the error rate of the PDU process and q the residual SDU or packet error rate when the RLC layer is not fully reliable. Figure 10.5 plots q against ε for a typical segmentation value of $K = 14$ and for $L = 1, 2, 3$ and 4. The residual packet error rate q is significantly higher than the equivalent residual PDU error rate $\varepsilon_{[res}$. For $\varepsilon = 0.1$, the residual packet error rate q is approximately 10 times greater than the equivalent value of residual PDU error rate ε_{res}. As ε increases towards unity, the ratio between q and ε_{res} also approaches unity. This is intuitive, since, when $\varepsilon_{res} \approx 1$ and almost all PDUs are erroneous, almost all SDUs or packets will be erroneous as well.

Figure 10.6 compares the iid case with three different values of channel burstiness, $b_\varepsilon = 10, 20$ and 30. Whereas the throughput of selective-repeat ARQ is insensitive to burst errors regardless of the window size m, Figure 10.6 shows clearly that the residual packet error rate q is affected by channel burstiness. One may observe from Figure 10.6 that an iid channel results in an upper bound for $\varepsilon < 0.2$ and a useful lower bound when $\varepsilon > 0.3$. As ε increases, the residual packet error rate q for $b_\varepsilon = 10$ is smaller than that for $b_\varepsilon = 30$. The same behaviour applies to the curve for $b_\varepsilon = 20$, which has a lower residual packet error rate than for $b_\varepsilon = 30$ when $\varepsilon > 0.5$. We may therefore conclude from Figure 10.6 that the channel burstiness has a diminishing impact on the residual

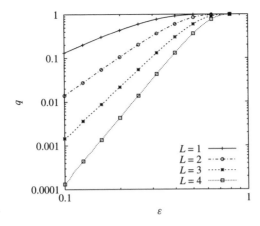

Figure 10.5   Residual packet error rate as a function of ε for different values of L

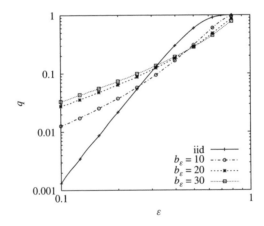

Figure 10.6   Residual packet error rate as a function of ε for different values of b_ε

packet error for increasing values of PDU error rate, whereas the opposite applies for small values of ε, such as $\varepsilon < 0.2$ in Figure 10.6.

Figure 10.7 plots the residual packet error rate as a function of PDU burstiness for $\varepsilon = 0.1$ and $L = 1$, 2 and 3. The values of q vary widely as a function of L, especially for $b_\varepsilon < 10$. For $b_\varepsilon > 10$, the curves for the three different values of L converge towards a value of $q \approx 0.4$.

Figure 10.8 depicts the average burst length b_q of the residual packet errors against the average burst length b_ε of the underlying PDU error process for $\varepsilon = 0.1$, 0.2 and 0.3. It can be seen that the average burst length of the residual packet errors for $b_\varepsilon = 100$ is 3 times that of $b_\varepsilon = 1$ (or approximately iid). All three curves converge towards $b_q \approx 3.6$ as b_ε approaches 100.

In summary, we have seen for the residual packet error rates of PTP channels that the burstiness of the underlying PDU process does not have a significant effect on the

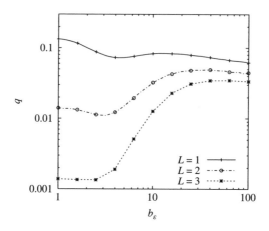

Figure 10.7 Residual packet error rate as a function of ε for different values of L

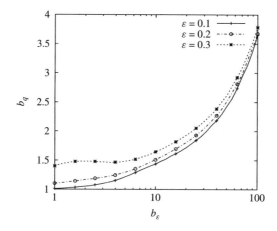

Figure 10.8 Residual packet loss burst length against b_ε for different values of ε

burstiness of residual packet errors, and that a small maximum allowable number of packet transmissions, for instance in order to keep packet delays low, does render the packet stream unreliable.

Packet Delay

We consider the delivery delay of an aggregate of $K = 14$ PDUs constituting a single SDU. The SDU delivery delay is defined as the number of slots elapsed between the slot where the first PDU of a given RLC SDU is transmitted for the first time over the channel and the slot where the receiver correctly receives the last PDU composing the SDU. We assume that SDUs are delivered in sequence to upper layers. Figure 10.9 plots the mean SDU delivery delay against the PDU error rate ε for $m = 10, 20, 30$ and 50. Figure 10.9 shows an upward shift of the mean SDU delivery delay owing to an increase in m. The

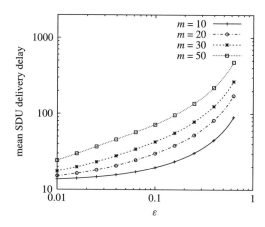

Figure 10.9 Mean packet delivery delay against ε for different values of m

SDU delivery delays for out-of-sequence delivery are only slightly smaller compared with those shown in Figure 10.9 for in-sequence delivery. Out-of-sequence delivery does not result in significant gains towards reducing packet delays.

10.4.2 PTM Channels for Multicast

We now consider the use of a single PTM channel for multicasting to a group of receivers.

Multicast Throughput

Proactively introducing parity packets to a TG is a simple way to reduce the packet error rate. Figures 10.10 and 10.11 plot the throughput improvement of HARQ1 compared with conventional ARQ without FEC against the number of receivers in the group, N_m. With conventional ARQ, a PDU is considered erroneous in a given slot and subsequently retransmitted if at least one multicast user does not correctly receive it (Rossi, Fitzek and Zorzi, 2003). Figure 10.10 considers a TG of size $k = 10$ with $h = 2, 4$ and 5. For $N_m < 3$, conventional ARQ without FEC has the highest throughput. The throughput curve of conventional ARQ, however, falls off quickly with increasing N_m. On the other hand, adding a small amount of redundancy (for example, $h = 2$) is sufficient when $N_m < 10$ but not when $N_m > 10$, since in that case HARQ1 only marginally outperforms ARQ without FEC.

In Figure 10.11, we consider a TG of size $k = 100$. As can be seen from the figure, $h = 10$ does not provide sufficient redundancy to ensure efficient throughput since HARQ1 with $h = 10$ has approximately the same throughput as ARQ without FEC. With $h = 20$, however, the throughput of HARQ1 is much improved compared with the case for $h = 10$ for $N_m < 100$ and gradually decreases for $N_m > 100$. With $h = 30$, on the other hand, the throughput of HARQ1 remains constant at $\tau = 0.76$. Having large TGs with HARQ1 leads to improved throughput performance, but also has drawbacks with respect to the delivery delay of individual packets. Especially over the air interface, where delivery delays should be kept to a minimum, small TG sizes are therefore preferable.

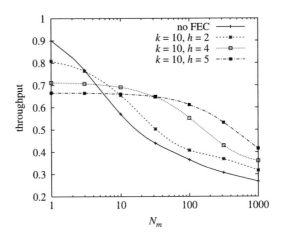

Figure 10.10   Multicast throughput of HARQ1 as a function of N_m with $k = 10$

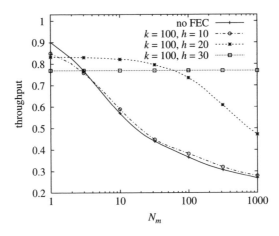

Figure 10.11   Multicast throughput of HARQ1 as a function of N_m for $k = 100$

Whereas proactively adding redundant parity packets reduces the packet error rate, transmitting parity packets instead of original packets as a result of packet losses has the advantage that only the required number of parities are transmitted, which improves throughput. Figure 10.12 plots the number of multicast users against the throughput of HARQ2 for different TG sizes, $k = 10$, 50 and 100. Clearly, large TGs result in improved throughput. As will be shown shortly, large TGs come at the cost of introducing large packet delivery delays, and as such do not always ensure a good trade-off between throughput and delay.

How parity retransmissions are distributed when the channel is bursty plays an important role in determining how efficient multicast throughput can be achieved. In Figure 10.13, we compare the throughput of HARQ2, HARQ3 and HARQ4 for a fixed group size of $N_m = 100$ for b_ε in the range [1:100]. Transmitting a single parity packet in response to

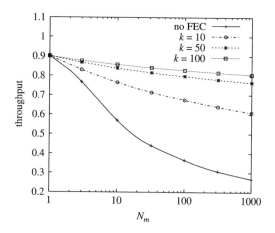

Figure 10.12   Multicast throughput of HARQ1 as a function of N_m

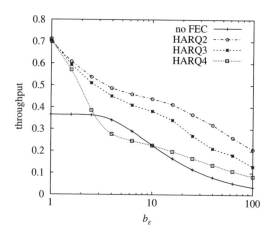

Figure 10.13   Comparison of ARQ without FEC, HARQ2, HARQ3 and HARQ4

packet errors, as with HARQ2, is the optimal way to achieve efficient throughput when long packet error bursts are frequent. HARQ4, on the other hand, which transmits parity packets immediately after the data packets of the TG, is the least efficient, since in this case parity packets are not sufficiently distributed over time in order to overcome long packet error bursts. A compromise solution that achieves a good trade-off between HARQ2 and HARQ4 is HARQ3, which transmits a variable amount of redundancy, depending on the maximum number of packet losses in a TG for all receivers.

As a final result, we consider a small to moderate group size of multicast users and a high packet error probability. Figure 10.14 plots the throughput against N_m for $\varepsilon = 0.85$. The figure clearly shows that, when the packet loss probability ε approaches unity, the use of parity-based loss recovery achieves a lower throughput than conventional ARQ without FEC.

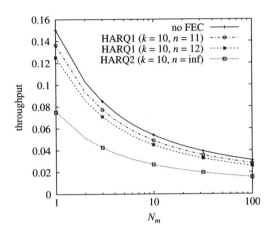

Figure 10.14   Comparison of ARQ without FEC, HARQ1 and HARQ2, with $\varepsilon = 0.85$

In conclusion, adding more parity either proactively or in response to packet errors results in a lower throughput than when no parity at all is introduced and only original data packets are retransmitted.

Residual Packet Error Rate

We now investigate parity-based loss recovery for reliable multicast when the number of retransmission attempts for each data packet and the number of additional parity packets at the sender are limited.

The iid case provided an upper bound on the residual packet error rate q with a single receiver for $\varepsilon > 0.3$, as shown in Figure 10.6. Considering a group of $N_m = 100$ receivers and $L = 2$, Figure 10.15 shows that, with HARQ1 and $k = 10$, this is not the case when $\varepsilon = 0.1$. Figure 10.15 shows that the channel burstiness has a negligible impact on the

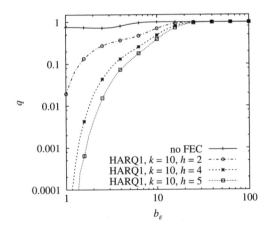

Figure 10.15   Residual packet error rate against b_ε for HARQ1, with $k = 10$

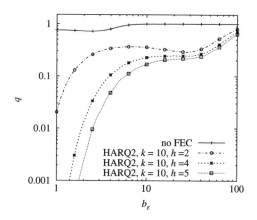

Figure 10.16   Residual packet error rate against b_ε for HARQ2, with $k = 10$

residual packet error rate when no parity-based loss recovery is used. On the contrary, we have observed with a range of different values for PDU error rate and channel burstiness that the iid case underestimates the residual packet error rate when FEC is employed.

When an infinite number of parity packets are available at the transmitter, the residual packet error rate is very small, such that $q = 0$. In order to estimate the residual packet error rate when the number of retransmission attempts and the number of available parity packets at the sender are limited, Figure 10.16 provides an upper bound (worst case) for HARQ2 with $k = 10$ and $L = 2$ for different values of h. Figure 10.16 plots q against the average channel burst length b_ε for $N_m = 100$. In comparing Figure 10.15 and 10.16, we observe that the residual packet error rates for both HARQ1 and HARQ2 are initially very similar for $b_\varepsilon < 5$, but with HARQ2 having a significantly lower residual packet error rate for $b_\varepsilon > 10$, as much as half that of HARQ1.

Figures 10.17 and 10.18 report the residual packet error rate and the residual packet error burst length of HARQ2 with $k = 10$ and $L = 2$. We take $\varepsilon = 0.1$ and $b_\varepsilon = 10$ as

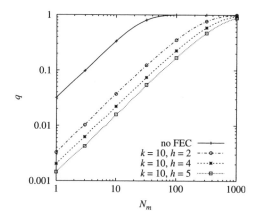

Figure 10.17   Residual packet error rate with $k = 10$

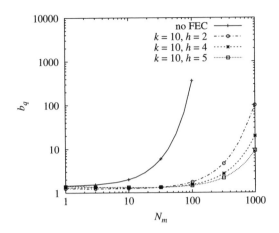

Figure 10.18   Average residual packet loss burst length against N_m for HARQ2

parameters of the PDU error process in Figures 10.17 and 10.18. The residual packet error rate in Figure 10.17 is logarithmic in N_m for $N_m < 400$ and approaches $q = 1$ as $N_m \approx 1000$. With ARQ, the residual packet error rate quickly approaches $q = 1$ for moderately large groups, for example $N_m > 100$.

Figure 10.18 plots b_q, the average burst length of the residual packet errors, for HARQ2 as a function of the multicast group size. With ARQ, the burst length increases very rapidly with N_m. Since $q = 1$ for $N_m > 100$ in Figure 10.17, the average burst length of the residual packet errors with ARQ is only shown for $N_m < 100$. For HARQ2, the average burst length of the residual packet errors is $b_q < 2$ and has a maximum of $b_q = 100$ with HARQ2 for $N_m = 1000$ when $h = 2$.

Overall, an important result is that very small residual packet error rates can be achieved for a single PTM channel with H-ARQ when an unlimited number of parity packets are available at the sender.

Packet Delay

We now show results for the packet delay with H-ARQ, for which we take $m = 50$. Figure 10.19 shows the mean SDU delivery delay for ARQ without FEC and for HARQ1 with $k = 10$ and $h = 4$ and 5. The SDU delivery delay for ARQ without FEC is logarithmic in N_m and approximately equal to $72 + 22 \ln(N_m)$, as shown in the figure. This means that the SDU delivery delay tends to infinity as N_m increases (Rossi, Zorgi and Fitzek, 2004). The SDU delivery delays for $h = 4$ and $h = 5$ are very similar for $N_m < 10$, with the gap between the two curves widening for $N_m < 300$ and then finally converging as $N_m = 1000$. Overall, HARQ1 results in a reduced SDU delivery delay compared with ARQ.

Figure 10.20 highlights how large TG sizes result in larger packet delivery delays by depicting the mean SDU delivery delays for ARQ without FEC and for HARQ1 with $k = 10$ and $h = 5$, $k = 50$ and $h = 25$ and $k = 100$ and $h = 50$. As expected, the SDU delivery delays for small TG sizes such as $k = 10$ are generally significantly smaller than for large TG sizes. Interestingly, for $N_m > 100$, HARQ1 with $k = 10$ has a longer delay

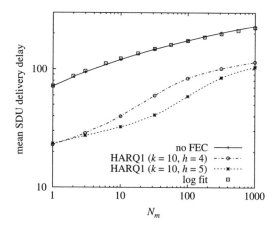

Figure 10.19   Mean packet delivery delay for HARQ1, with $k = 10$

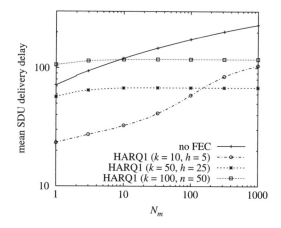

Figure 10.20   Mean packet delivery delay for HARQ1, with $k = 10$, 50 and 100

than HARQ1 with $k = 50$. The overall repair efficiency of HARQ1 with large TG sizes improves the delay performance as N_m increases.

Figure 10.21 plots the packet delivery delay as a function of ε for HARQ1, HARQ2 and HARQ3. For HARQ2 and HARQ3, we consider the proactive transmission of parity packets in order to reduce the delivery delay. For HARQ1 in Figure 10.21 we have $k = 10$ and $h = 5$, whereas for HARQ2 and HARQ3 we have $k = 10$ and $a = 3$. Initially, for $\varepsilon < 0.02$, the delivery delays for the three H-ARQ schemes are very similar, whereas the delivery delay for pure ARQ is significantly higher. For larger ε, the delay for HARQ1 remains initially constant, whereas the delivery delays of HARQ2 and HARQ3 increase with ε. For $\varepsilon > 0.1$, HARQ1 has the shortest packet delivery delay. The delays of HARQ1 and HARQ3 for $\varepsilon > 0.3$ are very similar, with HARQ2 having the longest overall delay. Whereas transmitting parity packets in response to packet losses is efficient in terms of throughput, this also comes at the cost of longer delivery delays. Figure 10.21 therefore

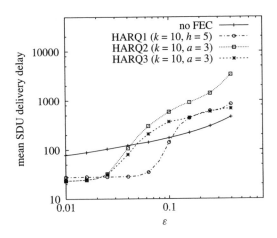

Figure 10.21   Mean packet delivery delay of HARQ1, HARQ2 and HARQ3

shows that H-ARQ may not always be optimal for applications with stringent latency requirements.

In summary, H-ARQ increases the packet delivery delay and therefore should be limited to a small number of parity retransmissions if packet delays are the key performance indicator, as, for instance, is the case with multimedia transmissions.

10.5   End-to-End Reliable Multicast

In MBMS, end-to-end reliability between the source and all receivers is provided by means of Raptor coding on the application layer. This mechanism ensures that packet errors within the wired part of the network and over the air interface can be repaired.

In contrast to MBMS, which does not rely on a back-channel but instead implements a file repair procedure at the end of the multicast transmission in order to repair packet losses, we consider a generic protocol for application-layer reliable multicast, in which acknowledgement messages are sent back to the transmitter (ACKs or NACKs) and the source performs retransmissions based on the acknowledgements. We assume that the feedback messages do not overwhelm the sender and feedback implosion does not occur as a result of these.

In this section, we investigate what effect the use of PTP or PTM channels for reliable multicast has on the reliability mechanism operating on the application layer. We evaluate the mean number of transmissions per received packet of the end-to-end reliability mechanism, both when PTP and when PTM channels are used. Based on the statistics of the residual IP packet or SDU errors in terms of error rate and average burst length obtained in the previous section, we evaluate the IP packet retransmission process of the reliable multicast mechanism operating on the application layer between the source and all N_m receivers. A detailed description of the simulation set-up is given in Section 10.3. For the residual packet errors of the RLC layer, we take the parameters of the residual packet error rate to be those shown in Figures 10.17 and 10.18 for the PTP channel and those shown in Figures 10.17 and 10.18 for the PTM channel configured for HARQ2.

These figures were plotted assuming a PDU error process with $\varepsilon = 0.1$ and $b_\varepsilon = 10$, and a maximum number of PDU retransmissions of $L = 2$.

In comparing the PTP and PTM schemes for reliable multicast, we consider only the last leg of transmission, the link over the air interface. Residual packet errors of the RLC layer trigger packet retransmissions or the transmission of additional parities on the application layer, depending on the configuration of the parity-based loss recovery mechanism.

Figure 10.22 compares the mean number of transmissions per received packet for the PTP and PTM schemes without FEC and with HARQ1 for $k = 10$ and $h = 1$. Our first observation is that the use of HARQ1 is not justified, as ARQ without FEC performs better than HARQ1. In comparing the PTP and PTM schemes without FEC, we find that the PTM scheme requires fewer transmissions for a single packet to be received by all receivers than the PTP scheme for $N_m < 300$, whereas the opposite is true for $N_m > 300$,

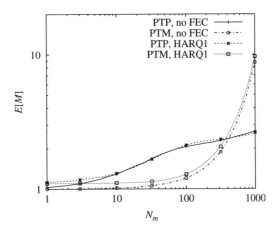

Figure 10.22 $E[M]$ against N_m for PTM and PTP schemes with HARQ1

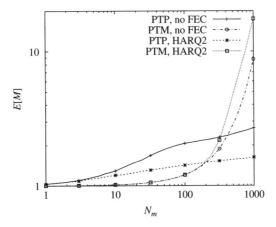

Figure 10.23 $E[M]$ against N_m for PTM and PTP schemes with HARQ2

in which case $E[M]$ increases rapidly for the PTM scheme. This is a result of the residual channel burstiness, which increases quickly for $N_m > 300$, as shown in Figure 10.18.

Figure 10.23 plots a similar result to Figure 10.22 but considering the performance of the PTP and PTM schemes for HARQ2 with $k = 10$. The use of HARQ2 for the PTP scheme is justified since it performs significantly better that plain ARQ for $N_m > 10$. With the PTM scheme, on the other hand, HARQ2 and ARQ without FEC perform nearly identically for $N_m < 100$. Similarly to the case with HARQ1 in Figure 10.22, pure ARQ outperforms HARQ2 when $N_m > 200$.

Figures 10.24 and 10.25 depict the packet delivery delays for the same set of parameters as the previous two graphs. The conclusions we can draw from Figures 10.24 and 10.25 are very similar to those from Figures 10.22 and 10.23. Both figures show that the use of H-ARQ is not justified since the packet delivery delay with ARQ without FEC is shorter

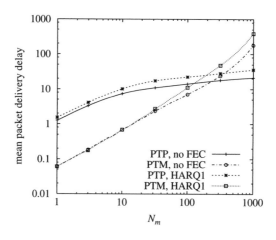

Figure 10.24   Packet delay against N_m for PTM and PTP schemes with HARQ1

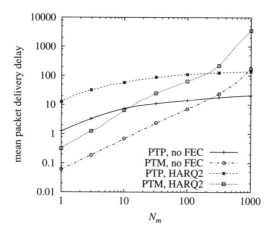

Figure 10.25   Packet delay against N_m for PTM and PTP schemes with HARQ2

than with H-ARQ. As with our previous findings, the PTM scheme generally has shorter delays when the number of receivers is moderate, whereas the PTP scheme results in lower latency for $N_m \approx 300$.

In conclusion, using a single PTM channel for reliable multicast results in fewer retransmissions for the reliable multicast protocol operating above the RLC layer for group sizes that are reasonably small. The opposite is true for large groups, where using separate PTP channels for individual users results in fewer retransmissions of the end-to-end reliability mechanism.

10.6 Summary

We have performed a simulation study of reliable multicast with and without the use of FEC. In our study, we have considered reliable multicast delivery over the air interface as well as on an end-to-end basis between the sender and all receivers. The performance metrics we have taken into account are the channel efficiency (in terms of throughput or the mean number of transmissions per received packet), delay and residual packet error rate when the RLC layer is configured not to be fully reliable.

Overall, H-ARQ by means of parity-based loss recovery improves the throughput of the multicast application significantly. In considering reliable multicast over the air interface, parity-based loss recovery is best performed with small TG sizes, since large TG sizes introduce large delivery delays. When the packet error is large, however, plain ARQ without the use of FEC achieves a higher throughput than H-ARQ.

Distributing the transmission of additional parity packets in response to packet errors within a TG improves the overall throughput when using a single PTM channel for multicast. This may be achieved by only transmitting a single additional parity packet when receivers experience packet losses. However, this comes at the cost of increased packet delays. Furthermore, H-ARQ can achieve very small residual packet error rates for the PTM channel used for multicast over the air interface when the number of parity packets available at the sender is not limited. Also, packet delays can be reduced if parity packets are transmitted proactively with each TG.

In comparing the use of PTP and PTM channels for multicast, we observed that a single PTM channel is preferable if the number of multicast users is reasonably small, and that separate PTP channels are the favourable solution with group sizes of $N_m = 1000$.

11

Mobile Multicast in Heterogeneous Networks

11.1 Introduction

Most of the content presented so far has focused on multicast in third-generation networks. This chapter looks beyond multicast in third-generation networks by providing an overview of alternative technologies for mobile multicast service delivery that fall outside the traditional mobile network domain. Also, several trends that may result as a consequence of a heterogeneous network landscape are presented.

The first part of this chapter is concerned with alternative technologies for mobile multicast delivery. Traditionally, the delivery of content to a large receiver audience has been the domain of broadcast technologies. It is therefore not surprising that a variety of solutions for mobile multicast delivery based on digitial broadcasting have emerged in recent years. Many of the systems are evolutions of existing digital broadcast standards, which have been enabled to transport IP-based data traffic and are optimized for efficient mobile reception by handheld receiver terminals. Examples of such evolving standards are DVB-H or Terrestrial Digital Multimedia Broadcast (T-DMB). Other technologies such as MediaFLO, on the other hand, have been designed from scratch for efficient mobile multicast reception, without issues of backwards compatibility.

The second part of this chapter covers the topic of multicast delivery in a heterogeneous networking environment. It is expected that several multicast-capable network technologies will coexist in future network environments. These technologies can complement each other in terms of functionality and delivery characteristics, coverage, radio and network resources. Combining such technologies allows network operators to significantly enhance their service delivery to mobile users in terms of delivery quality and service availability. Also, novel value-enhanced services may be offered by combining network technologies. Much research has been done in recent years on enabling efficient interworking and convergence of heterogeneous network technologies for unicast service delivery. Achieving efficient delivery of multicast services in such a heterogeneous environment poses additional challenges owing to the fact that the mechansims must scale well for a

Multicast in Third-Generation Mobile Networks   Robert Rümmler, Alexander Gluhak and A. Hamid Aghvami
© 2009 John Wiley & Sons, Ltd

potentially large group of receivers. The concept of convergence and interworking in the context of different multicast-capable wireless networks is therefore briefly explored, and challenges specific to the delivery of multicast services are highlighted. The chapter closes with a possible solution for the efficient delivery of multicast services in a heterogeneous network environment.

11.2   Alternative Technologies for Mobile Multicast

In this section, we introduce several alternative technologies for multicast to mobile receivers. The alternative technologies we consider are DVB-H, MediaFLO, ISDB-T and DMB. Whereas the WiMAX standard may be appropriate for mobile multicast delivery in larger geographic areas, it is not treated in this book.

11.2.1   DVB-H

DVB-H (ETSI, 2007) is one of the latest standards in the family of digital broadcasting standards developed by the DVB project (DVB Project, 2008). As indicated by its name, DVB-H addresses the digital transmission of IP-based broadcasting content to mobile battery-powered and pocket-sized devices such as mobile phones and Personal Digital Assistants (PDAs). The DVB-H standard is based on DVB-T, the digital broadcasting standard for terrestrial reception. However, DVB-H extends DVB-T by the introduction of new features that minimize the power consumption at the receiver and improve the mobile reception of broadcast content.

DVB-H solutions typically operate in the Very High-Frequency (VHF) and Ultrahigh-Frequency (UHF) bands of terrestrial television and can make use of channels of 6, 7 and 8 MHz bandwidth. They can also utilize 5 MHz channels outside the classical broadcast bands. DVB-H is designed to be operated on top of DVB-T transmission networks, making use of the same network infrastructure and using part of the multiplex capacity used for conventional broadcast programmes. This is particularly appealing for broadcast operators. Furthermore, dedicated DVB-H networks can be deployed utilizing entire multiplexes for DVB-H transmission. Such networks are able to take full advantage of all DVB-H features (Faria et al., 2006). Mobile network operators, for example, could deploy DVB-H networks, significantly enhancing their existing network capacity for delivering broadcast and multicast content to their mobile subscribers. Transmitters for DVB-H can cover up to several tens of kilometres and provide up to 13 Mbps of bandwidth per UHF channel multiplex.

A possible realization of a DVB-H network is depicted in Figure 11.1. Broadcasting content such as video and audio and additional data services are often encoded in national or regional broadcasting centres and transported as MPEG-2 (where MPEG stands for Moving Pictures Expert Group) transport streams via a (private) broadcast distribution network to the DVB-H transmitter sites in the country. The broadcast distribution network can be multicast-enabled in order to support the transport of IP multicast data more efficiently. The management of services and IP Encapsulation (IPE) is usually controlled from the broadcasting centres. The lower part of Figure 11.1 shows the internals of a IP/DVB gateway, in which MPEG-2 video and audio services are broadcast together with IP-based data services over the same Transport Stream Multiplex (TS-MUX). In such a scenario, a part of the multiplex is reserved for IP-based data, which is inserted into the transport stream via IPE. The upper part of Figure 11.1 assumes the case where a multiplex is used entirely for the transfer of IP-based data.

Figure 11.2 provides an overview of the DVB-H protocol stack. DVB-H (ETSI, 2004) actually acts as an umbrella standard encompassing several specifications that address

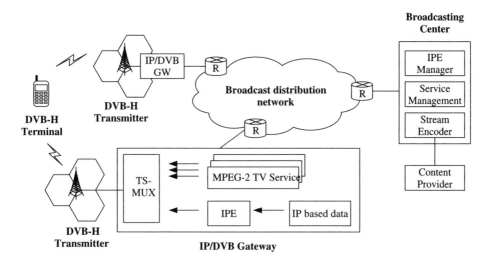

Figure 11.1   Example of a DVB-H network

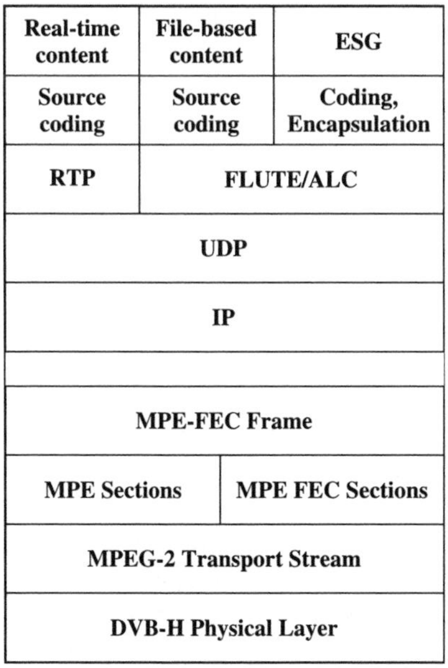

Real-time content	File-based content	ESG
Source coding	Source coding	Coding, Encapsulation
RTP	FLUTE/ALC	
UDP		
IP		
MPE-FEC Frame		
MPE Sections	MPE FEC Sections	
MPEG-2 Transport Stream		
DVB-H Physical Layer		

Figure 11.2   The DVB-H protocol stack

different aspects of the depicted protocol stack. The lower layers of the protocol stack operate as defined by extensions to existing DVB specifications, while the higher layers operate as defined by the IP datacast specification (ETSI, 2005). Both are described in more detail in the following.

The Lower Layers of DVB-H

The DVB-H physical layer is based on the DVB-T specification. The transmission parameter signalling channel has been enhanced to advertize the availablity of features specific to DVB-H for the elementary transport streams carried in a multiplex. The cell identifier, which is optional in DVB-T, is mandatory in DVB-H networks as it supports the discovery of neighbouring network cells and thus enables mobility across cell boundaries. In addition, the DVB-H physical layer defines an additional OFDM transmission mode, the so-called 4k mode. This mode offers greater flexibility in network planning (Kornfeld and May, 2007).

For DVB systems, MPEG-2 has been chosen as the standard for high-quality compression of digital video streams. MPEG-2 also defines a transport mechanism to multiplex various video, audio and data streams together to one common Transport Stream (TS), allowing for synchronized play-out of all media components at the receiver. Compressed media and data components are packetized into fixed-sized MPEG-2 TS packets of 188 bytes, which are combined to logical channels. Several logical channels are transmitted within one TS.

The logical channels are identified by 13-bit Packet Identifiers (PIDs) which are part of the header MPEG-2 TS packets. Theoretically, up to 8192 channels can be transmitted inside one TS. However, a small number of PID channels are reserved for service-specific signalling. This additional signalling information is generated in the form of service information tables, allowing the receiver to tune to a specific channel within the transport stream and to reassemble and decode the original video, audio or data stream.

The DVB project has specified a data broadcasting standard (ETSI, 2003) that allows various data types such as IP data to be included in an MPEG-2 TS. Multi-Protocol Encapsulation (MPE) has been recommended for the encapsulation of IP data packets in a TS. Incoming IP data packets are first encapsulated in MPE packets, then segmented to fit into the MPEG-2 TS packets and finally multiplexed onto a logical channel of the TS. Every MPE packet includes the next-hop destination MAC address in its header.

Each logical channel is identified by a PID. The receiver is able to identify the channel carrying IP data by consulting the service information tables that are generated during IP encapsulation. Since the number of logical channels in one TS is limited, packets for different data connections can be multiplexed into one channel. Packets belonging to the same data stream are uniquely identified by the PID, the source address and the destination address and are referred to as an IP stream. In order to allow a receiver to locate an IP stream in the TS sent by the DVB transmitter, additional service information needs to be generated during encapsulation. For this purpose, an IP MAC notification table is used to signal the availability and location of an IP stream in a TS.

The DVB-H data link layer introduces several new features on top of the MPEG-2 TS layer that make DVB-H reception more power efficient and capable of supporting mobility across different broadcast cells. Conventional DVB-T requires receivers continously to listen to the transmitted data stream in order to access a given service of interest.

Continous operation of the receiver front-end is a serious drawback for battery-powered receiver devices. The DVB-H standard introduces a new technique referred to as *time slicing*. With time slicing, service multiplexing is performed in a time-division multiplex. The service data is not transmitted continuously but in compact periodical bursts with longer interruptions in between. Based on additional signalling provided, a receiver can time-selectively listen to only the burst carrying the service data of interest and turn off the receiver front-end for periods in which data for other services is transmitted. This technique is similar to discontinuous reception in MBMS and can result in power savings at the receiver exceeding 90 %. Time slicing also allows receivers during the power-save period to search for channels in neighbouring broadcast cells that offer the same selected service. Monitoring of neighbouring cells during the power-save periods does not disrupt ongoing reception and allows seamless handovers between two cells to be performed. In order to increase the robustness of the transmitted signal, the physical-layer FEC of the the DVB-T standard is complemented by error coding on the link layer. This results in significant coding gains at the receiver. This technique is referred to as MPE-FEC and is applied to IP input streams prior to MPE encapsulation. The scheme consists of a combination of Reed-Solomon coding with extensive time interleaving with flexible configuration options.

IP Datacast

IP datacast, short for IP-based data broadcast, refers to a set of specifications that provide the upper layers of DVB-H with the ability to support IP-based data transport. The IP datacast specifications also define a mechanism to incorporate an optional interactive network, providing a return link to a DVB-H-based broadcast network for interactive service access. Similarly to MBMS, the IP datacast standard provides delivery protocols suitable for both real-time streaming delivery and reliable download delivery of IP multicast data. For streaming delivery, RTP is utilized on top of a UDP/IP. MPEG-4 H.264 is recommended as the coding format for video streaming, with optional support for the VC-1 codec used in Microsoft Windows Media 9. Audio content is encoded with MPEG-4 AAC+ (where AAC stands for Advanced Audio Coding). Reliable download delivery of file-based content is based on the FLUTE protocol. Repair of file segments that have been corrupted during transmission is achieved either through repeated transmission of the file as part of a data carousel or through selective retransmissions, in case a return channel through an interactive network is available.

In order to assist users in discovering and selecting multicast and broadcast service content that may be offered in a DVB-H network, the IP datacast standard defines the Electronic Service Guide (ESG). The service guide carries meta-information describing the services available and providing schedule information as well as other parameters necessary for the selection of a service programme. The information for ESG is transferred reliably using the FLUTE protocol.

Compared with the service information tables used at the lower levels to recover MPEG-2 programmes or IP data streams, the ESG offers a much richer set of features to advertise services to users. Supported features include the ability to provide hyperlinks, pictures and video trailers. The ESG specification also provides a multilayer model efficiently to represent and encode the meta-information that constitutes the ESG. Further details on DVB-H can be found in (Faria *et al.*, 2006) and (Kornfeld and May, 2007).

11.2.2   MediaFLO

MediaFLO, with FLO standing for Forward Link Only, is a unidirectional transmission system optimized for the efficient distribution of multimedia content to a large group of mobile receivers (Qualcomm, 2007). Unlike MBMS or DVB-H, which are global standards, MediaFLO is a proprietary standard developed by Qualcomm.

Similarly to DVB-H, MediaFLO operates in the UHF, VHF and L bands in channels of 5, 6, 7 and 8 MHz bandwidth. In contrast to DVB-H, MediaFLO has been designed completely from scratch and as such is not compatible with any other standards. This has enabled the system to be designed in a manner that has been optimized in terms of spectral efficiency and power consumption at the receiver (Chari *et al.*, 2007).

As depicted in Figure 11.3, the MediaFLO system consists of FLO transmitters that are connected to a backhaul network and FLO-enabled receiver devices. The network and the content to be provided are managed by national and local network operation centres, facilitating the dissemination of content with both nationwide and regional relevance to FLO-enabled receivers. An uplink for service subscription and security key distribution can be realized with an existing cellular network.

An overview of the protocol stack is provided in Figure 11.4. The physical layer of MediaFLO is based on OFDM and supports only the 4k mode, which provides a sound trade-off between mobility and cell size. It incorporates FEC techniques consisting of an inner Turbo code and an outer Reed-Solomon error-correcting code. A logical channel structure in the MAC layer has been designed that allows for different coding and modulation schemes, thus supporting various reliability and QoS requirements. The variable-rate codecs also allow the system to obtain statistical multiplexing gains when serving real-time content at variable rates over different logical channels. For the delivery of real-time content, MediaFLO deliberately makes use of optimized non-IP-based streaming solutions, eliminating the additional overhead of IP-based protocols. Non-real-time content such as text or graphics is delivered over an IP-based protocol stack. More details on the protocols can be found in the standard documents (TIA, 2006; TIA, 2007).

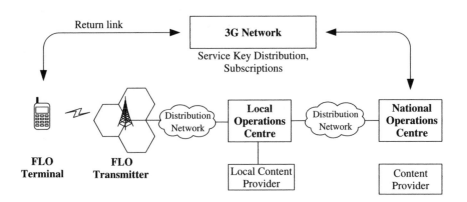

Figure 11.3   Overview of a MediaFLO network

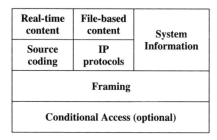

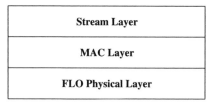

Figure 11.4   The MediaFLO protocol stack

11.2.3   Other Standards

Apart from DVB-H and MediaFLO, several other technologies are available that are suitable for mobile reception of multicast and broadcast services. Two standards worth mentioning are the Japanese ISDB and the South Korean DMB standards.

ISDB-T was developed in the late 1990s in Japan and is widely deployed there for digital broadcast of terrestrial television programmes. ISDB-T corresponds to the DVB-T standard and is an OFDM system that operates channels of 6 MHz bandwidth in the VHF and UHF range. Like DVB-T, IDSB-T makes use of the MPEG-2 transport stream and corresponding codecs for the transmission of video and audio programmes. One of the key differences to DVB-T, however, is that ISDB-T uses a modulation scheme referred to as segmented transmission OFDM. This modulation scheme splits a transmission band into 14 equal segments. Thirteen of these OFDM segments can be used for transmission of broadcast services, with the transmission parameters for each segment being individually configured (Takada and Saito, 2006). Based on the transmission configuration, a receiver can selectively receive segments of the signal, without requiring decoding and reception of the complete signal transmitted over the whole bandwidth of the channel. This partial reception allows for significant power savings at the receiver, thus making it suitable for portable devices. A typical configuration would make use of 12 segments for digital terrestrial television, while one segment, the centre segment, can be used for the reception of audio or data with an approximate bandwidth of 430 kHz.

DMB (ETSI, 2007) is another standard that allows for reception of multimedia content by portable devices and mobile phones. T-DMB, the terrestrial version, is a narrowband solution requiring channels of 1.5 MHz in the VHF/UHF frequency bands. It was developed as an evolution of the European Digital Audio Broadcasting (DAB) standard (ETSI, 2001) and was mainly driven in South Korea. While DAB is most suitable for the transfer of digital audio content with low-rate data services for digital radio reception, T-DMB has adopted the technoloy for the delivery of video streams together with audio and data services. Efficient source coding based on the MPEG-4 standard enables a significant reduction in the required data rate for multimedia content. The resulting lower bit error rate requirements are met with additional channel coding schemes based on Reed-Solomon codes and byte-interleaving schemes. In addition, T-DMB allows for the provisioning of data service with a variety of protocol options including IP-based data delivery. Today, T-DMB is commercially deployed in several countries, with South Korea, China and Germany being early adopters. Compared with DVB-H or MediaFlo, DMB has a fairly low entry barrier in terms of deployment costs, resulting from lower costs of transmission equipment and the fact that fewer transmitter sites are required to provide adequate coverage. Another advantage of DMB lies in the advantageous availability of spectrum in some countries. In contrast to DVB-H or MediaFlo, a T-DMB multiplex has a limited capacity with a maximum transmission bandwidth of about 1.1 Mbps, thus limiting the number of services an operator can provide. The initial cost advantage of T-DMB may quickly fade if a larger number of services need to be offered.

11.3   Interworking and Convergence

The availability of different multicast-capable wireless network technologies provides new opportunities for network operators. Networks consisting of different wireless access technologies can complement each other in terms of functionality and delivery characteristics, coverage, radio and network resources. The concept of cooperation between different heterogeneous network technologies is one of the fundamental concepts envisioned in the landscape beyond third-generation mobile communications.

For example, unidirectional networks such as DVB-H can easily benefit from complementary functionality of alternative access networks such as UMTS, for instance by providing a missing return link for a user interaction channel. Without a return link, the delivery of multicast or broadcast content is based on a predefined schedule, which needs to be configured in advance by the network operator or content provider. The availability of a return channel allows users dynamically to request the delivery of a particular broadcast service. A broadcast operator is thus able to optimize and dynamically adapt the service schedule and use the network resources more efficiently by providing only desired or sufficiently popular services. An interaction channel realized by a return link also allows the operator to offer more enhanced services to their customers. By embedding links into their broadcast content, interactive services with the users can be realized, for instance by providing additional information to a particular broadcast content or by receiving feedback from the user through polls and voting. Likewise, by using broadcast network technology such as DVB-H, a mobile network operator can provide higher data rate services to its mobile customers, beyond the bearer limits of the mobile network. Increased network capacity through the use of alternative radio resources and spectrum is another example of the advantages of joint utilization of different multicast-capable wireless network technologies.

Recent years have seen significant research work on enabling the integration or interworking of different wireless network technologies, with an initial focus on unicast communications (3GPP, 2007a). The levels of cooperation may vary significantly. Network cooperation can be achieved by simply providing a single bill for the use of two different networks, providing security credentials for access to different integrated networks or even allowing access to network-internal services of a home network from another network. More advanced mechanisms allow forms of cooperation in which all services may be accessed from one or the other network or even seamlessly transferred between different network technologies without the user experiencing disruption of a service. Such a transfer is referred to as a vertical handover. New network technology can be attached directly to the radio access network (referred to as integration), connected to the core network (referred to as tight coupling) or interworked via external network gateways (referred to as loose coupling). Tighter coupling and integration offer an advantage for performance-critical cooperation, for example when performing a vertical handover of a real-time service, as they introduce lower latencies for interworking mechanisms compared with loose-coupling approaches.

There are two basic operational scenarios for the cooperation of heterogeneous networks. The first scenario assumes that a network operator expands its existing network with an alternative wireless network technology administered underneath a single authority. Such a scenario typically enables a tighter form of cooperation and improved performance for service delivery between wireless network technologies. However, this solution typically has large investment costs as it requires operators to acquire new transceiver sites, network equipment and possibly licences for new spectrum.

In the second scenario, an operator is expected to cooperate with an operator of an alternative network technology while both operators retain administrative autonomy and control of their networks. This form of cooperation, also referred to as network interworking (Tuttlebee *et al.*, 2003), allows operators with different access networks to enhance their coverage and capacity for delivering services, without the need to acquire licenses for additional spectrum or purchase new costly access network infrastructure. In contrast to the first scenario, only looser forms of coupling can be achieved.

The next section introduces the challenges of network cooperation, with a focus on multicast service delivery. Subsequently, an approach for multicast service delivery in a heterogeneous wireless network environment is presented and discussed.

11.4   Challenges for Multicast Delivery in Heterogeneous Networks

A major challenge in a heterogeneous network environment is the lack of delivery coordination owing to the purely receiver-driven service model of IP multicast. In order to establish a multicast session, a mobile receiver enables the reception of data from a particular multicast group on one of its interfaces and sends a join request to the attached access network.

Figure 11.5 illustrates the problem using a simple example. All receivers have different network access interfaces available, but they are assumed to be at the same location and expected to receive the same flow of session data. Receivers 1 and 2 join the service on the UMTS network, receivers 3 and 4 choose to initiate the session via WLAN and receiver 5 requests the multicast service on DVB. Consequently, the same service content is delivered via three different data paths to the same location. Knowing that all the receivers have a UMTS network access interface, the receivers could instead be instructed to join the UMTS network for receiving the traffic. Another challenge is the inherent heterogeneity

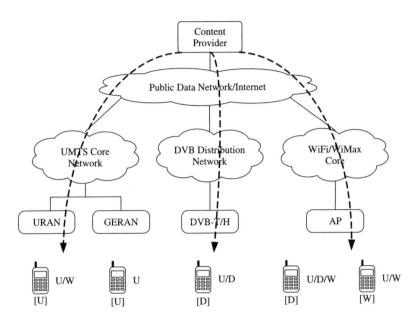

Figure 11.5   Multicast service delivery in heterogeneous networks

of mobile receivers, which often makes it impossible to satisfy all interested receivers with a single service flow within a multicast session or with a common access network for the delivery. The provision of appropriate service flows to satisfy all interested receivers in a multicast session and the selection of suitable delivery networks require knowledge of the existing heterogeneity of the receiver population and available networks. Advanced resource and session management mechanisms would allow for multicast service delivery with higher coordination.

11.5   Multicast Delivery Coordination in Heterogeneous Networks

In this section, an interworking architecture is presented that allows different operators to offer cooperative services to users across a heterogeneous wireless network environment, both securely and seamlessly. As described above, there is a need for additional network control in a heterogeneous network environment. This requires a shift in the paradigm from a purely receiver-driven to a network-controlled provision of multicast services. Cooperating network operators deploy so-called Interworking Gateways (IGWs) within their networks. An IGW implements various essential interworking functions and defines a logical link among them, which enables signalling message exchanges for interworking purposes. The IGW ensures that operators may interwork efficiently for service delivery, without the need to give up their own network autonomy and to disclose any sensitive network-related information. Figure 11.6 provides an overview of the interworking architecture. The main parts of the proposed solution for coordinated delivery of multicast services are distributed resource management functionality and context-aware group management support. Both are described in the following.

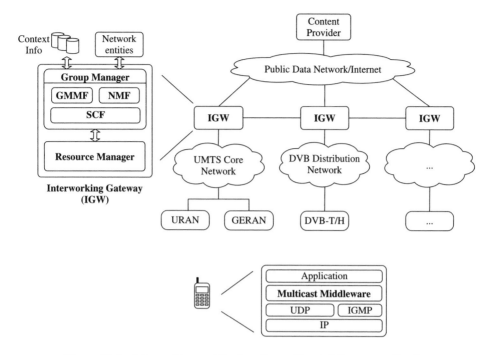

Figure 11.6 Interworking architecture for multicast delivery coordination

11.5.1 Resource Management

Resource management support is responsible for the efficient utilization of network resources for multicast service delivery within the access networks of different cooperating operators.

Current resource management mechanisms in mobile networks assume that the network resources are under the control of a single network operator. Resource management decisions are therefore often handled by centralized entities. The reason for this is that network operators want to retain full control of their respective network resources and are reluctant to disclose control or even traffic load information to entities outside their administrative domains. In order to address this problem, a distributed resource management framework has been developed.

Several resource management techniques can be used by resource management algorithms to optimize the delivery of multicast services. The resource management techniques mainly targeted at multimedia multicast services are service scheduling, delivery network selection and service QoS selection. *Service scheduling* is used by multimedia-on-demand services to determine the optimal service set-up time for a group of users that are batched as a multicast group. *Delivery network selection* determines a suitable network for the delivery of the multicast service for a specific user. *Service QoS selection* adapts the service transmission QoS according to resource availability. This is done by selecting a certain service flow with a suitable service quality for a receiver, given that multiple service flows at different qualities are provided.

Different algorithms can be used with respect to each of the above resource management techniques. These algorithms aim to solve two problems introduced in the context of multicast service provisioning over heterogeneous wireless networks, namely the service scheduling problem (that is, when to transmit the required service) and the user assignment problem (that is, which network a user is connected to and what QoS a user is served with). The optimal selection of resource management algorithms depends on the policy defined by multicast service providers or operators.

11.5.2 Group Management Support

In the current multicast service model, users that are interested in receiving data of a multicast service subscribe to a multicast group address on a particular network interface. Thus, a user selects not only the desired multicast service but also implicitly the multicast bearer service and the network for the delivery. Furthermore, an application for the multicast service at the receiver will usually maintain this association for the whole duration of a session. In order to overcome this lack of flexibility, a decoupling of the multicast user service and its service flows from multicast bearer services in the networks is required.

The group management support provides necessary mechanisms to achieve the required decoupling. Instead of directly subscribing to a multicast group on the network layer, receivers notify their interest in receiving a multicast user service by subscribing to a network-layer-independent group at the group manager for that multicast service. Considering resource management decisions, a session of a multicast service can thus be dynamically instantiated in different service flows and network connections to reflect the heterogeneity of receivers and available access networks at their location.

As shown in Figure 11.6, the group management support consists of a context-aware group manager entity on the network side and a multicast middleware on the terminal side. Furthermore, for scalable delivery of required control signalling between the group manager and the multicast middleware in the terminals, a multicast signalling channel is utilized. Details of these entities and their contribution to achieve the desired multicast delivery coordination are presented in the following.

Group Manager

Group management support plays two significant roles in the provision of multicast services. Firstly, it accumulates and maintains knowledge of currently interested multicast receivers for a multicast service and assists the resource manager in its decisions by providing relevant context information of these users as well as other service-related information. Secondly, it provides the necessary session management functionality for delivery coordination by executing the resource management decisions. The functionality is realized by three functional building blocks, namely the Group Membership Management Function (GMMF), the Session Control Function (SCF) and the Network Management Function (NMF). These three functions are briefly described below.

The GMMF is concerned with the collection and maintenance of the group membership for a multicast service. It provides a set of functional primitives that allow the creation and management of groups as well as service-related parameters. The functional primitives can be separated into primitives used by service providers to configure a multicast service at the group manager and primitives used by mobile users to register their interest for a particular service. The SCF provides the session management functionality required for multicast delivery coordination on the network side. The task of the SCF is to implement the decisions of the Resource Manager (RM). It realizes the control plane mechanism on the session layer by managing the network-layer service at the receivers through the multicast middleware. The NMF deals with the network-side configuration, which may be required for multicast bearer management in the access networks. MBMS, for instance, requires the configuration of multicast bearer service parameters in BM-SC before multicast bearer services can be established. The NMF provides the required service parameters to the BM-SC, before the SCF triggers the establishment of the respective multicast bearer services at the receivers. Likewise, the NMF configures adequate DVB/IP gateways or gateway managers in the DVB network.

Multicast Signalling Channel

The introduction of a control plane for the coordination of multicast data delivery adds a signalling load to the networks. One way of delivering the control signalling is by sending a signalling message separately to each receiver. However, with a potentially large number of users, the required signalling load in the network may increase significantly. Therefore, scalable delivery of such signalling is critical in order to keep the impact of the additional control plane as low as possible.

A scenario-based analysis has revealed that a control signalling message will often target a larger subset of receivers. As a consequence, a multicast signalling channel can be used. The multicast signalling channel enables scalable delivery of control-plane signalling for large groups of receivers. For each multicast user service, a separate multicast signalling channel is utilized for the control plane. After user subscription to a multicast service at the group manager, the multicast middleware starts listening to the respective multicast signalling channel in order to receive notifications from the group manager.

Multicast Middleware

The group management support on the user terminal is realized by a multicast middleware, which sits directly underneath the application layer, leaving the lower transport and network layer stack of the mobile terminal unchanged. The purpose of the middleware is to manage the network-layer services on the terminal as indicated by the SCF. Furthermore, it hides changes in the underlying network connection to the application throughout the lifetime of a multicast session. Multicast socket calls of the application are intercepted by the middleware, which in turn handles the actual network socket operations.

Once a user chooses to subscribe to a multicast service, the multicast middleware starts listening to a multicast signalling channel that was indicated by the respective service announcement. After the subscription, the GM in the network takes control of the multicast session and *steers* the middleware in the terminals via the multicast signalling channel. Based on the notifications, the middleware layer opens or closes required multicast network sockets on respective interfaces, establishing or releasing the indicated multicast bearers and passing received session data to the application. The basic operations of the multicast middleware are depicted in Figure 11.7. The following service example illustrates the steps in more detail and provides an overview of the necessary signalling involved between the entities of the group management support.

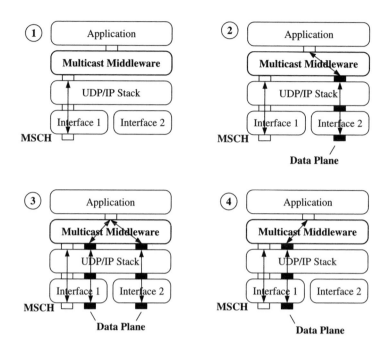

Figure 11.7 Multicast middleware hides the establishment of different bearer services from the application

11.5.3 Service Example

Figure 11.8 shows an example of a network for network-initiated bearer establishment at the beginning of a service session. Users learn about a service session and its relevant configuration parameters, for example by visiting a portal website. If users are interested, they register for a multicast user service with the GMMF by providing the group ID for the service and a user identifier. The GMMF adds the user to the group membership database of the service and obtains the relevant context information of the user. Meanwhile, the GMMF triggers the SS function in the resource manager with user and service-related context information. Consequently, user requests are batched in different batching queues based on the characteristics of their required services. Users also obtain the Multicast Group Address (MGA) of the multicast signalling channel for the service and possible signalling networks from the service announcement and provide this information to the multicast middleware. The multicast middleware initiates the establishment of the multi-cast bearer service for the multicast signalling channel by sending an IGMP join message to the AR of the signalling network. An AR can, for example, be a multicast-enabled router behind a WLAN access point or part of a GGSN in case of a UMTS network. IGMP signalling itself may not be sufficient for multicast bearer establishment. In an UMTS network, for example, an IGMP join would trigger the additional execution of an MBMS multicast service activation procedure in order to establish a multicast bearer. Once the service session is scheduled to start by the SS function, the RM selects suitable service flows and delivery networks for the interested group of receivers and triggers the

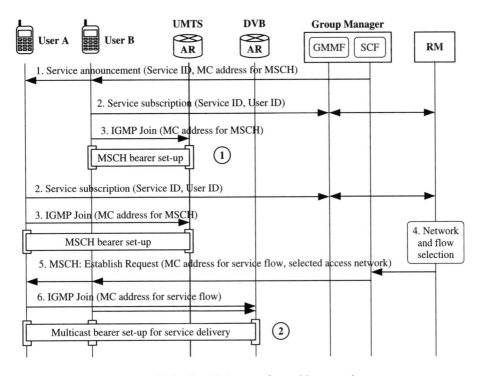

Figure 11.8   Establishment of a multicast session

SCF to initiate the establishment of the required multicast bearers for the data plane. The SCF determines the receiver subset for the selected flows and delivery networks. It then sends an appropriate ESTABLISH message for the identified receiver subsets, providing MGA and other parameters required to establish the data plane for the session. The multicast middleware in the user terminals then joins the multicast group on identified networks to establish required bearer services and begins forwarding incoming service data to the application.

Figure 11.9 depicts an example of a vertical network handover for a receiver group. A vertical handover can happen, for instance, for load-balancing reasons. The RM reacts to resource conditions in the network and triggers the SCF to perform a vertical network handover, balancing a subgroup of receivers in parts of the network to an alternative access network. The SCF identifies the target receiver subset and sends a MIGRATE message on the multicast signalling channel. The message identifies the MGA of the multicast bearer service and the old access network from which and the new access network to which the bearer is migrating. The multicast middleware at the identified receiver terminals executes the command by sending IGMP join requests to the new access network to establish the new multicast bearer service. Once data is received on the new bearer, the multicast middleware in the terminals sends an IGMP leave request on the old network. As a consequence, the previous multicast bearer service is released. The corresponding actions at the middleware for both establishment and migration of the bearers can be observed in Figure 11.7.

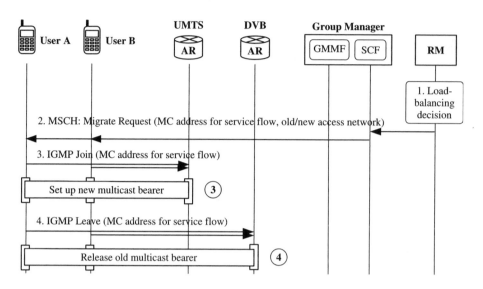

Figure 11.9   Example of a vertical handover of a group of users for load-balancing reasons

11.6   Summary

This final chapter has taken a look at the delivery of mobile multicast in digital broadcast networks. The two most prominent broadcasting standards are the DVB-H and MediaFLO. DVB-H is an evolution of the terrestrial digital broadcasting standard DVB-T, adding further improvements that make it suitable for mobile reception with battery-powered handheld devices. The robustness of mobile reception is improved by a combination of new physical layer options and efficient error coding at the MAC layer, while significant savings in power consumption can be achieved with efficient service multiplexing and the introduction of time-slicing techniques. While DVB-H is backed by the DVB project, a large multinational alliance, MediaFLO has emerged as a proprietary alternative that, in contrast to the DVB-H standard, does not aim to be compatible with terrestrial broadcasting standards. Only time will tell which of the two technologies provides the right balance between performance, cost and adequate vendor support and will thus excel in the market.

The second part of the chapter considered multicast service provision in a heterogeneous network environment. The benefits of cooperation of different mobile network technologies were briefly explored, considering the cases of a converged network environment under a single operator as well as the case of interworking operators employing different network technologies. The challenges for mobile multicast delivery in such an environment were introduced. The chapter concluded with an example of an interworking solution that achieves multicast delivery coordination in a heterogenous network environment.

Appendix A

Derivation of Downlink Capacity

In this appendix, closed-form expressions for the capacity of the PTP and PTM channel selection schemes for multicast over the CDMA air interface are derived.

A.1 Ratio of Intercell Interference to Received Power

Calculating $z_i(r_i, \theta_i)$ for all possible values of (r_i, θ_i) is difficult. Owing to the symmetry of the three-tier cell layout shown in Figure A.1, the worst-case I_{oc}/S_0 is found at the six corners of the reference cell, at which the numerical values for I_{oc}/S_0 are equal. In order to establish a reference case for the interference that is analytically tractable, we calculate I_{oc}/S_0 at a distance d from the home BS with $\theta = 0$ (that is, along the x-axis superimposed on Figure A.1) and take this as a worst case in terms of interference (Choi and Kim, 2001a). Doing so is justified firstly because $S_0 = P_0 \cdot L_0$ is independent of θ and therefore only a function of d, and secondly because evaluating I_{oc} at an arbitrary distance d with $\theta = 0$ takes into account the dominant sources of interference from the neighbouring cells. The I_{oc}/S_0 at a distance d from the home BS with $\theta = 0$ is thus taken as the worst-case interference power for any MS located at (d, θ).

The received power from the home BS at a distance d from the home BS is $S_0 = P_0 \cdot L_0 = d^{-\omega} \cdot \chi_0$. As in (Choi and Kim, 2001a), we only consider the intercell interference from the 11 closest BSs, since the interference from any of the other BSs is negligible compared with that of the 11 closest BSs. With the cell numbering as displayed in Figure A.1, and assuming that all BSs are transmitting at full power, the ratio of intercell interference power to received power from the home BS as a function of distance d from the home BS with $\theta = 0$, written as $z(d) = I_{oc}/S_0$, is given by

$$
z(d) = \frac{I_{oc}}{S_0} = \sum_{k=1}^{11} \frac{P_k \cdot L_k}{P_0 \cdot L_0} = \sum_{k=1}^{11} \left(\frac{D_k}{D_0} \right)^{-\omega} \cdot \frac{\chi_k}{\chi_0}
$$

$$
= \frac{(3-d)^{-\omega} \chi_1 + [(1.5-d)^2 + 0.75]^{-\omega/2} \sum_{k=2}^{3} \chi_k}{d^{-\omega} \chi_0}
$$

Multicast in Third-Generation Mobile Networks Robert Rümmler, Alexander Gluhak and A. Hamid Aghvami
© 2009 John Wiley & Sons, Ltd

$$+ \frac{[(3-d)^2+9]^{-\omega/2} \sum_{k=4}^{5} \chi_k + [(1.5-d)^2+6.75]^{-\omega/2} \sum_{k=6}^{7} \chi_k}{d^{-\omega} \chi_0} \qquad (A.1)$$

$$+ \frac{(d^2+3)^{-\omega/2} \sum_{k=8}^{9} \chi_k + [(d+1.5)^2+0.75]^{-\omega} \sum_{k=10}^{11} \chi_k}{d^{-\omega} \chi_0}.$$

Let m_i and σ_i denote the mean and standard deviation (in dB) of I_{oc}, and m_s and σ_s the mean and standard deviation (in dB) of S_0 respectively. The mean and standard deviation of I_{oc} and S_0 may be calculated from the numerator and denominator of Equation (A.1) respectively. The mean m_z and standard deviation σ_z of I_{oc}/S_0 (in dB) may then be calculated as $m_z = m_i - m_s$ and $\sigma_z = \sqrt{\sigma_i^2 + \sigma_s^2}$ respectively.

We evaluate the sum of lognormally distributed random variables, such as in the numerator of Equation (A.1), using Ho's method (Ho and Akyildiz, 1996), which provides better accuracy in calculating the mean and standard deviation than other popular methods (Cardieri and Rappaport, 2001).

As the MS moves from the centre of the cell, the total received power from the home BS S_0 decreases, whereas the intercell interference I_{oc} increases. As a result of this, the mean m_z and standard deviation σ_z of I_{oc}/S_0 increase as the MS moves towards and beyond the cell boundary. With a shadowing spread of $\sigma = 4.5$ dB, and in using Ho's method for calculating the intercell interference, the mean and standard deviation of I_{oc}/S_0 (in dB) at a distance d from the home BS are tabulated in Table A.1.

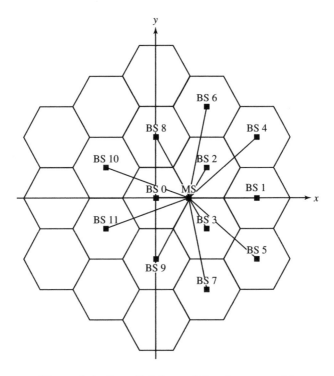

Figure A.1 Downlink intercell interference model

Table A.1 Mean and standard deviation of I_{oc}/S_0

d	m_z (dB)	σ_z (dB)
1.00	4.98	5.27
1.025	5.58	5.28
1.05	6.17	5.29
1.075	6.75	5.3
1.10	7.31	5.31

A.2 Derivation of Average Downlink Power Factor

The average downlink power factor η was first derived by Lee and Miller (1999) for a single-cell system and subsequently by Choi and Kim (2001a) for a multicell system. Our derivation differs from that of Choi and Kim (2001a) in that we take into account the variation of I_{oc} over the cell boundary, which is more appropriate for a multicell system.

Let $z(r, \theta)$ denote the I_{oc}/S_0 at a specific location (r, θ) in the cell. Since we are interested in determining the variation of I_{oc}/S_0 within the cell as a function of location, we ignore shadow fading and express I_{oc}/S_0 only as a function of distance-dependent propagation loss. The distance of the jth BS at location (R_j, θ_j) in polar coordinates from an MS in the home cell at the location (x, y) (in polar coordinates) is defined as $d_j = \sqrt{R_j^2 + x^2 - 2R_j x \cos(|\theta_j - y|)}$.

The value of I_{oc}/S_0 averaged over the entire cell area is calculated as

$$\overline{z(x, y)} = \int_0^{\theta_s} \int_0^{R_c} x \left(\frac{\sqrt{R_j^2 + x^2 - 2R_j x \cos(|\theta_j - y|)}}{x} \right)^{-\omega} f_{r,\theta}(x, y)\, dx\, dy, \quad \text{(A.2)}$$

where ω is the path loss exponent and $f_{r,\theta}(x, y)$ is the probability density function (pdf) of all MSs within the sector, and θ_s is the sectorization angle. The pdf $f_{r,\theta}(x, y)$ is given by

$$f_{r,\theta}(x, y) = \begin{cases} 2/R_c^2 \theta_s, & 0 \le x \le R_c,\ 0 \le y \le \theta_s, \\ 0, & \text{otherwise.} \end{cases}$$

Since we consider an unsectorized cell, $\theta_s = 2\pi$ and the pdf becomes

$$f_{r,\theta}(x, y) = \begin{cases} 1/\pi R_c^2, & 0 \le x \le R_c,\ 0 \le y \le \theta_s, \\ 0, & \text{otherwise.} \end{cases}$$

Now we calculate I_{oc}/S_0 averaged over the circumference of the cell boundary, in other words at a distance R_c from the BS:

$$\overline{z(R_c, y)} = \int_0^{\theta_s} R_c \left(\frac{\sqrt{R_j^2 + R_c^2 - 2R_j R_c \cos(|\theta_j - y|)}}{R_c} \right)^{-\omega} f_{r,\theta}(R_c, y)\, dy, \quad \text{(A.3)}$$

where

$$
f_{r,\theta}(R_c, y) = \begin{cases} 2\pi R_c, & 0 \le y \le \theta_s, \\ 0, & \text{otherwise,} \end{cases}
$$

Here, the pdf reflects the worst-case scenario and represents the distribution of MSs along the circumference of the home cell. In (Choi and Kim, 2001a), the derivation of η only considers a single point on the cell boundary as the worst case. This is not appropriate in a multicell system, as I_{oc} varies along the circumference of the home cell. The average downlink power factor η may now be expressed as $\eta = \overline{z(x, y)}/z(R_c, y)$, where $\overline{z(x, y)}$ and $z(R_c, y)$ are given by Equation (A.2) and Equation (A.3) respectively. With $\omega = 4$ as the path loss exponent, the average downlink power factor evaluates to $\eta = 0.24$.

A.3   Multicast Capacity with PTP Channels

As derived in the previous section, assuming a uniform distribution of users within the cell, and with $\omega = 4$ as the path loss exponent, the average I_{oc}/S_0 of the ith user is $\eta = 0.24$ times that of the averaged worst-case at the cell boundary. This allows us to express the average transmission power allocated to the ith user as

$$
\overline{\phi_i} = \eta \cdot \alpha_i \cdot \frac{\gamma \cdot R}{W \psi} \cdot y(R_c),
$$

where α_i is the duty cycle or activity factor of the ith user. The time-averaged transmission power is therefore reduced by the factor α_i compared with the case for continuous transmission, for which $\alpha_i = 1$.

The received E_b/I_0 at the MS is different according to the data rate R and the relative fraction of power ϕ_i allocated to the ith user. The received E_b/I_0 for voice and multicast users is

$$
\gamma_v \approx \frac{W \cdot \psi}{R_v} \cdot \frac{\phi_i^{(v)}}{z_i(r_i, \theta_i) + \varphi} = \frac{W \cdot \psi}{R_v} \cdot \frac{\phi_i^{(v)}}{y_i(r_i, \theta_i)} \tag{A.4}
$$

$$
\gamma_m \approx \frac{W \cdot \psi}{R_m} \cdot \frac{\phi_i^{(m)}}{z_i(r_i, \theta_i) + \varphi} = \frac{W \cdot \psi}{R_m} \cdot \frac{\phi_i^{(m)}}{y_i(r_i, \theta_i)}, \tag{A.5}
$$

where R_v and R_m are the data rate, γ_v and γ_m are the required E_b/I_0 and $\phi_i^{(v)}$ and $\phi_i^{(m)}$ are the fractions of BS transmit power allocated to the ith user for voice and multicast services respectively. After rearranging Equation (A.4) and Equation (A.5), the received fraction of power allocated to the ith user for the voice and multicast service may be written as

$$
\phi_i^{(v)} = \frac{\gamma_v \cdot R_v}{W \psi} \cdot y_i(r_i, \theta_i) \tag{A.6}
$$

$$
\phi_i^{(m)} = \frac{\gamma_m \cdot R_m}{W \psi} \cdot y_i(r_i, \theta_i). \tag{A.7}
$$

When power control is not ideal, the error is assumed to be lognormally distributed with standard deviation σ_e (in dB) (Tam and Lau, 1999). If the PCE is considered, another multiplicative lognormal factor is to be introduced. As a result of PCE, the mean and standard deviation of $y_i(r_i, \theta_i)$ in Equation (A.6) and Equation (A.7) are $m_t = m_y$ and $\sigma_t = \sqrt{\sigma_y^2 + \sigma_e^2}$, where m_y and σ_y are the respective mean and standard deviation of $y_i(r_i, \theta_i)$ without PCE. The PCE thus only affects the standard deviation of $y_i(r_i, \theta_i)$, whereas its mean remains unchanged.

On the downlink, all common and dedicated channels share the limited total transmission power of the BS. Outage therefore occurs when the total BS power is not sufficient to satisfy the power requirements of all users. By taking into account the average downlink power factor η, the BS outage probability is

$$
P_b = \Pr\left[\eta \left(\sum_{i=1}^{K_v} \alpha_i^{(v)} \cdot \phi_i^{(v)} + \sum_{i=1}^{K_m} \alpha_i^{(m)} \cdot \phi_i^{(m)} \right) > 1 \right]
$$

$$
= \Pr\left[Z_v + Z_m = Z > 1 \right],
$$

(A.8)

where K_v and K_m are the number of voice and multicast data users and $\alpha_i^{(v)}$ and $\alpha_i^{(m)}$ are the activity factors or duty cycles for individual voice users and multicast users respectively. We approximate the terms $\eta \sum_{i=1}^{K_v} \alpha_i^{(v)} \phi_i^{(v)}$ and $\eta \sum_{i=1}^{K_m} \alpha_i^{(m)} \phi_i^{(m)}$ by lognormal random variables Z_v and Z_m respectively, since $\sum_{i=1}^{K_v} \phi_i^{(v)}$ and $\sum_{i=1}^{K_m} \phi_i^{(m)}$ are independent and identically distributed lognormal random variables. Since $\tilde{Z} = 10\log(Z) = \log(Z_v + Z_m)$ is a normal random variable, the BS outage probability may be written as (Choi and Kim, 2001a)

$$
P_b = \Pr\left[\tilde{Z} = 10\log(Z) > 0 \right]
$$

$$
= Q\left(\frac{-E[\tilde{Z}]}{\sqrt{Var[\tilde{Z}]}} \right),
$$

where $Q(x) = \frac{1}{2\pi} \int_x^\infty e^{-x^2/2}\, dx$.

A common assumption is that each cell of the network behaves like an independent $M/M/\infty$ queue (Evans and Everitt, 1999). Consequently, the number of voice and multicast users, K_v and K_m, are Poisson random variables with mean $\rho_v \lambda_v / \mu_v (1 + g)$ and $\rho_m \lambda_m / \mu_m (1 + g)$ respectively. Here, $\lambda_{(\cdot)}$ and $\mu_{(\cdot)}$ are the arrival and departure rates for the voice and multicast services respectively. Under the assumption that the traffic to every cell in the network is equal (which applies to the parameters of a stochastic model and is distinctly different to assuming an equal static load in every cell (Evans and Everitt, 1999)), g is the fraction of users from neighbouring cells that are in soft handover. The users that are in soft handover contribute approximately twice as much interference compared with users that are not in soft handover (Viterbi, 1995). We assume that a fraction $g < 1$ of all users are in a two-way soft handover, and that both BSs involved in the soft handover allocate essentially the same fraction of power to the soft-handover users. The effective arrival rate per BS increases then from λ to $\lambda(1 + g)$ (Viterbi, 1995). The

Erlang capacity is the set of values $A_v = \lambda_v/\mu_v$ and $A_m = \lambda_m/\mu_m$ that keep the BS outage probability at a target level, typically 0.01 (Choi and Kim, 2001a).

A.4   Multicast Capacity with PTM Channels

Let us assume that the hard handover occurs at a reasonable distance beyond the cell boundary at a distance r_c from the home BS, with $r_c > R_c$ (Viterbi, 1995). The BS must therefore transmit at a power level that can be received at a distance r_c from the home BS, where $r_c > R_c$. The fraction of power required for reliable transmission to the ith user located at a distance r_c from the home BS is given by

$$\phi_i^{(c)} = \frac{\gamma_c R}{W \psi} \left[z(r_c) + \varphi \right]. \tag{A.9}$$

Let ϕ_c denote the fixed fraction of total BS transmission power allocated to the PTM channel. With hard handover, in the worst case ϕ_c must be sufficient to provide an MS at a distance $r_c > R_c$ from the home BS with a link that achieves the required performance all but a fraction P_c of the time, where P_c is the link outage probability (Viterbi, 1995). Here, the link outage probability is defined as the probability that an MS on the PTM channel achieves an insufficient E_b/I_0.

The link outage probability P_c may be written as

$$P_c = \Pr\left[Z_c = \phi_i^{(c)} > \phi_c \right]$$
$$= \Pr\left[\tilde{Z}_c = 10 \log \left(\phi_i^{(c)} \right) > 10 \log \left(\phi_c \right) \right],$$

where $\phi_i^{(c)}$ at a distance r_c from the BS is given by Equation (A.9). Since $\tilde{Z}_c = 10 \log(Z_c)$ is a normal random variable, the link outage probability is

$$P_c = Q\left(\frac{10 \log(\phi_c) - E[\tilde{Z}_c]}{\sqrt{Var[\tilde{Z}_c]}} \right). \tag{A.10}$$

We calculate ϕ_c such that Equation (A.10) is satisfied for a target-link outage probability P_c at a distance $r_c > R_c$ from the home BS.

The fraction of power ϕ_c allocated for the PTM channel is only broadcast by the BS when there is data to transmit. This leads to a reduction in the time-averaged transmission power for the PTM channel by the factor α_m, the duty cycle of the multicast source. Taking into account voice users, the BS outage probability may then be derived as

$$P_b = \Pr\left[\eta \sum_{i=1}^{K_v} \alpha_i^{(v)} \cdot \phi_i^{(v)} > 1 - \alpha_m \phi_c \right]$$
$$= Q\left(\frac{10 \log(1 - \alpha_m \phi_c) - E[\tilde{Z}_v]}{\sqrt{Var[\tilde{Z}_v]}} \right).$$

Appendix B

Cost Derivation of Multicast Routing

In this appendix, expressions for the cost of multicast packet delivery and location update in UMTS networks are derived. The packet delivery cost for broadcast, multiple unicast, dynamic multicast and MBMS is derived first. This is followed by a derivation of the location update cost of the standard UMTS mechanisms and that of MBMS.

B.1 State Probabilities

The notation for the state probabilities is given below:

P_{act}	probability that UE is SM active
P_{in}	probability that UE is SM inactive
P_{det}	probability that UE is PMM-DETACHED
P_{conn}	probability that a UE is PMM-CONNECTED
P_{RA}	probability that UE is PMM-IDLE
P_{URA}	probability that UE is URA connected
P_{cell}	probability that UE is cell connected

The RA information of subscribers is stored at the SGSN for PMM-IDLE subscribers. The location of a subscriber that is PMM-CONNECTED is known by the SGSN with the accuracy of the routing address of the actual serving RNC. Furthermore, a subscriber that is PMM-CONNECTED is tracked by the RNC on cell or URA level. The probability that a UE is PMM-CONNECTED is

$$P_{conn} = P_{cell} + P_{URA} \tag{B.1}$$

The probabilities of P_{det}, P_{RA} and P_{conn} sum to 1. The probability that a UE is PMM-DETACHED is therefore

$$P_{det} = 1 - P_{RA} - P_{conn}. \tag{B.2}$$

Multicast in Third-Generation Mobile Networks   Robert Rümmler, Alexander Gluhak and A. Hamid Aghvami
© 2009 John Wiley & Sons, Ltd

A user that is PMM-DETACHED is always SM inactive. An SM active user may either be PMM-IDLE or PMM-CONNECTED. Let P_a denote the probability that a subscriber that is PMM-IDLE or PMM-CONNECTED is SM active. Then the probability that a subscriber is SM inactive is given by

$$P_{in} = P_{det} + (1 - P_a) \cdot P_{RA} + (1 - P_a) \cdot P_{conn}.$$

Similarly, the probability that a user is SM active is

$$P_{act} = P_a \cdot P_{RA} + P_a \cdot P_{conn}.$$

We assume that all multicast users are always reachable by the network and thus either in the PMM-CONNECTED or PMM-IDLE state, and therefore $P_{det} = 0$.

B.2   Cost Variables

We assume that there is a cost associated with each link and node of the network, both for signalling and packet deliveries, such that the following notations apply.

D_{gs}	transmission cost of packet delivery between GGSN and SGSN
D_{sr}	transmission cost of packet delivery between SGSN and RNC
D_{rb}	transmission cost of packet delivery between RNC and Node B
D_{bu}	transmission cost of packet delivery over the air interface
S_{hg}	transmission cost of signalling between HLR and GGSN
S_{gs}	transmission cost of signalling between GGSN and SGSN
S_{sr}	transmission cost of signalling between SGSN and RNC
S_{rb}	transmission cost of signalling between RNC and Node B
S_{bu}	transmission cost of signalling over the air interface
p_g	processing cost of packet delivery at GGSN
p_s	processing cost of packet delivery at SGSN
p_r	processing cost of packet delivery at RNC
p_b	processing cost of packet delivery at Node B
a_h	processing cost of signalling at HLR
a_g	processing cost of signalling at GGSN
a_s	processing cost of signalling at SGSN
a_r	processing cost of signalling at RNC
a_b	processing cost of signalling at Node B

B.3   Packet Delivery Cost

In the following section, the packet delivery costs for broadcast, multiple unicast, dynamic multicast and MBMS are derived analytically.

Packet Delivery Cost – Broadcast

Firstly, we derive the cost of the simplest of the four one-to-many schemes, namely broadcasting multicast packets to all nodes within the network. In order to broadcast multicast packets, the network simply floods packets to all nodes within the network, without performing any paging. The total cost of multicast packet delivery by means of broadcasting can be derived as

$$D_I = \left[p_g + N_{S/G} \cdot (D_{gs} + p_s) + N_{RNC} \cdot (D_{sr} + p_r) \right. $$
$$\left. + N_B \cdot (D_{rb} + p_b + D_{bu}) \right] \cdot N_p \cdot \lambda_s, \tag{B.3}$$

where λ_s represents the multicast session arrival rate and N_p the number of packets transmitted during a single multicast session. It is evident from Equation (B.3) that the packet delivery cost of the broadcast scheme is independent of the number of multicast users, N_m.

Packet Delivery Cost – Multiple Unicast

One-to-many packet delivery can be performed by unicasting packets to all multicast members. This is achieved by duplicating a multicast packet at the GGSN for all group members and subsequently forwarding the duplicated packets to each group member separately. The total cost of multicast packet delivery with multiple unicast depends on the MM and RRC states of the multicast users. Firstly, we derive the cost of a single packet delivery for a multicast user that is cell connected. With a subscriber that is cell connected, neither the SGSN nor the RNC has to perform any paging. The cost of a packet delivery without any paging is given by

$$D_{cell} = p_g + D_{gs} + p_s + D_{sr} + p_r + D_{rb} + p_b + D_{bu}. \tag{B.4}$$

If a subscriber is URA connected, the RNC must first page all cells within the URA in which the subscriber is located before data transfer can take place. The RNC pages the subscriber by sending a paging message to all cells within the URA. The cell in which the subscriber is located then responds with the cell identity of the paged subscriber. The cost of paging a subscriber that is URA connected is

$$V_{URA} = N_{B/U} \cdot (S_{rb} + a_b + S_{bu}) + S_{bu} + a_b + S_{rb} + a_r. \tag{B.5}$$

If the MM state of a subscriber is PMM-IDLE, the SGSN only stores the identity of the RA in which the subscriber is located. Therefore, all cells in the RA must be paged in order to establish the routing address of the serving RNC. Once the SGSN has paged the subscriber, the location of the subscriber is known by the SGSN with the accuracy of the routing address of the serving RNC and by the RNC on cell level. The cost of paging a subscriber that is PMM-IDLE is

$$V_{RA} = N_{R/R} \cdot (S_{sr} + a_r) + N_{B/RA} \cdot (S_{rb} + a_b + S_{bu}) $$
$$+ S_{bu} + a_b + S_{rb} + a_r + S_{sr} + a_s, \tag{B.6}$$

where $N_{B/RA} = N_{R/R} \cdot N_{U/R} \cdot N_{B/U}$ is the total number of Node Bs per RA.

With the first packet of a data session, the SGSN must perform the paging procedure for a multicast user that is PMM-IDLE with probability P_{RA}. Similarly, the RNC must perform the paging procedure for a multicast user that is URA connected with probability P_{URA}. No paging is required for multicast users that are cell connected, and therefore packets may be transferred directly with probability P_{cell}. The total packet delivery cost for the multiple unicast scheme is derived as

$$
\begin{aligned}
D_{II} = \big[& P_{cell} \cdot D_{cell} \cdot N_p + P_{URA} \cdot (V_{URA} + D_{cell} \cdot N_p) \\
& + P_{RA} \cdot (V_{RA} + D_{cell} \cdot N_p) \big] \cdot N_m \cdot \lambda_s,
\end{aligned}
\tag{B.7}
$$

where V_{URA}, V_{RA} and N_m are defined according to Equations (B.5), (B.6) and (9.4) respectively.

Packet Delivery Cost – Dynamic Multicast

With our dynamic multicast scheme, multicast group management is performed at the GGSN, SGSN and RNC and multicast tunnels are established over the Gn and Iu interfaces. All multicast users that are in the PMM-IDLE state at the SGSN must be paged. After paging a multicast user that is PMM-IDLE, the RNC stores the location of the UE on cell level. The paging cost is given by Equation (B.6).

The manner in which multicast packets are transmitted within UTRAN depends on radio resource management and may take place on PTP or PTM channels (Holma and Toskala, 2007). We assume that the RNC forwards a single copy of each multicast packet only to those Node Bs that are serving multicast users. The Node Bs subsequently broadcast the multicast packet over the cell. The total cost of the multicast mechanism may be derived as

$$
\begin{aligned}
D_{III} = \big[& p_g + n_{S/G} \cdot (D_{gs} + p_s) + n_{RNC} \cdot (D_{sr} + p_r) \\
& + n_B \cdot (D_{rb} + p_b + D_{bu}) \big] \cdot N_p \cdot \lambda_s \\
& + (P_{RA} \cdot V_{RA} + P_{URA} \cdot V_{URA}) \cdot N_m \cdot \lambda_s.
\end{aligned}
\tag{B.8}
$$

Packet Delivery Cost – MBMS

Packet distribution in MBMS takes place according to the procedures described in (3GPP, 2007d, 3GPP 2008a). The packet delivery cost is equivalent to that of dynamic multicast, with the exception that in MBMS no paging is performed. Instead of paging subscribers, the session start and stop procedures inform potential recipients that a session is about to commence or to terminate. The cost of a session start or stop procedure (3GPP, 2007d) is

$$
\begin{aligned}
V_{IV} = 2 \cdot n_{S/G} \cdot (S_{gs} + a_s) + 2 \cdot n_{S/G} \cdot N_{R/S} \cdot (S_{sr} + a_r) \\
+ 2 \cdot n_{S/G} \cdot N_{R/S} \cdot N_{U/R} \cdot N_{B/U} \cdot (S_{rb} + a_b + S_{bu}) + N_m \cdot (S_{bu} + a_b + S_{rb} + a_r).
\end{aligned}
$$

The total cost of packet delivery for MBMS consists of packet distribution as well as the cost of the session start and stop procedures. It may be written as

$$
\begin{aligned}
D_{IV} = \big[& p_g + n_{S/G} \cdot (D_{gs} + p_s) + n_{RNC} \cdot (D_{sr} + p_r) \\
& + n_B \cdot (D_{rb} + p_b + D_{bu}) \big] \cdot N_p \cdot \lambda_s + 2 \cdot V_{IV} \cdot \lambda_s.
\end{aligned}
$$

B.4   Location Update Cost

In this section, we derive expressions for the location update costs. Broadcast does not require location update, although the standard location update procedures must still be performed in order to deliver regular traffic to subscribers. Multiple unicast relies on the standard location update procedures for mobility management. The additional location update procedures of our multicast scheme are negligible compared with the total cost of location updates. We therefore consider the location update costs of broadcast, multiple unicast and our multicast scheme to be the same. For the analysis, we consider a square-shaped network configuration, with the same notation for the cost variables as used previously.

Location Update Cost – Standard UMTS Mechanism

We now derive expressions for the standard location update procedures in UMTS. The standard location update procedures apply to broadcast, multiple unicast and our multicast scheme. The location update cost for MBMS will be derived in the next section.

A cell or URA update is a request-reply procedure between the subscriber and the serving RNC. Subscribers that are either cell or URA connected perform cell and URA updates respectively. The signalling messages for a cell or URA update are identical. The cost is

$$U_{cell} = U_{URA} = 2 \cdot (S_{bu} + a_b + S_{rb}) + a_r.$$

Here, the Node B must process two signalling messages, whereas the RNC processes only one.

We first consider SRNS relocation and RA update procedures that take place within the service area of an SGSN. The SRNS relocation is only performed for subscribers that are PMM-CONNECTED and always entails a cell or URA update. Drawing on the messages for the SRNS relocation depicted as defined in (3GPP, 2007b), the cost for an intra-SGSN SRNS relocation is given by

$$U_{RNC} = U_{cell} + 8 \cdot S_{sr} + 5 \cdot a_s + 4 \cdot a_r + S_{rr}.$$

An intra-SGSN RA update is a request-reply procedure between a subscriber and the serving SGSN. An intra-SGSN RA update is performed for both PMM-IDLE and PMM-CONNECTED subscribers. The cost of an intra-SGSN RA update for a PMM-IDLE subscriber is

$$U_{RA}^{(i)} = 2 \cdot (S_{bu} + a_b + S_{rb} + a_r + S_{sr}) + a_s.$$

In our system model of a square-shaped network configuration, an intra-SGSN RA update for a PMM-CONNECTED subscriber always results in an SRNS relocation. The cost of an intra-SGSN RA update for a subscriber that is PMM-CONNECTED is therefore

$$U_{RA}^{(c)} = U_{RNC} + U_{RA}^{(i)}.$$

The signalling messages for an inter-SGSN RA update are described in (Lin *et al.*, 2001). An inter-SGSN RA update is performed by first transferring the subscriber's point of attachment to the new SGSN and subsequently performing an RA update. As described in (Lin *et al.*, 2001), the cost for a PMM-IDLE subscriber is

$$U_{SGSN}^{(i)} = U_{RA}^{(i)} + 5 \cdot S_{rr} + 7 \cdot a_s + 2 \cdot S_{gs} + a_g$$
$$+ 4 \cdot S_{hs} + 3 \cdot a_h + S_{bu} + a_b + S_{rb} + a_r + S_{sr}.$$

PMM-CONNECTED subscribers moving between SGSNs must additionally perform an SRNS relocation. Therefore, the cost of an inter-SGSN RA update for a subscriber that is PMM-CONNECTED is

$$U_{SGSN}^{(c)} = U_{SGSN}^{(i)} + U_{RNC}.$$

In order to derive the cost of location updates per unit time, we combine the number of crossings for the different areas and the location update costs. The location update cost per unit time for subscribers that are PMM-IDLE is

$$U_I^{(i)} = P_{RA} \cdot (M_{RA} \cdot U_{RA}^{(i)} + M_{SGSN} \cdot U_{SGSN}^{(i)}).$$

The location update cost per unit time for subscribers that are PMM-CONNECTED is

$$U_I^{(c)} = P_{cell} \cdot M_B \cdot U_{cell} + P_{URA} \cdot M_{URA} \cdot U_{URA}$$
$$+ (1 - P_{RA}) \cdot (M_{RNC} \cdot U_{RNC} + M_{RA} \cdot U_{RA}^{(c)} + M_{SGSN} \cdot U_{SGSN}^{(c)}).$$

The total location update cost per unit time is the sum of these two and thus given by

$$U_I = U_I^{(c)} + U_I^{(i)}.$$

Location Update Cost – MBMS

With MBMS, each node maintains a list of downstream nodes that are serving multicast users and thus wish to receive multicast traffic addressed to a particular group. Each time the serving SGSN of a multicast user changes, the new SGSN must join the multicast tree by informing its parent GGSN to update its multicast routing table. Similarly, the old SGSN must leave the multicast tree by informing its parent GGSN that it wishes to leave the multicast tree. These two procedures are referred to as MBMS registration and deregistration respectively (3GPP, 2007d). In addition to updating the multicast tree, each SGSN must also ensure that its parent GGSN has a valid MBMS UE context for each group member. The signalling for registration, deregistration and MBMS UE context request/response procedures are performed independently (3GPP, 2007d). The cost of an MBMS registration/deregistration or MBMS UE context exchange between an SGSN and GGSN is

$$C_{gs} = 2 \cdot (S_{gs} + a_g) + a_s. \tag{B.9}$$

The RNC performs the RNC registration and deregistration procedures in order to make the CN aware that it is hosting multicast users (3GPP, 2008a). The cost of such a procedure is given by

$$C_{sr} = 2 \cdot (S_{sr} + a_s) + a_r. \tag{B.10}$$

In addition to the registration and deregistration of nodes, the UE linking procedure (3GPP, 2008a) allows one RNC to inform another that a multicast user is entering its service area. The associated cost is given by

$$C_{rr} = 2 \cdot (S_{rr} + a_r).$$

As depicted in 3GPP (2007d), with MBMS the update of the serving SGSN (that is, the inter-SGSN RA update) entails three separate signalling exchanges between the GGSN and the involved SGSNs. This results in the term $3 \cdot C_{gs}$ being added to the cost of an inter-SGSN RA update. The location update cost for PMM-IDLE users then becomes

$$U_{II}^{(i)} = P_{RA} \cdot \left[M_{RA} \cdot U_{RA}^{(i)} + M_{SGSN} \cdot (U_{SGSN}^{(i)} + 3 \cdot C_{gs}) \right].$$

With MBMS, an SRNC relocation entails the UE linking procedure between the involved RNCs, as well as MBMS registration/deregistration towards the serving SGSN. The location update cost for a subscriber that is PMM-CONNECTED is

$$\begin{aligned}
U_{II}^{(c)} = {} & P_{cell} \cdot M_B \cdot U_{cell} + P_{URA} \cdot M_{URA} \cdot U_{URA} \\
& + (1 - P_{RA}) \cdot [M_{RNC} \cdot (U_{RNC} + C_{rr} + 3 \cdot C_{sr}) \\
& + M_{RA} \cdot (U_{RA}^{(c)} + C_{rr} + 3 \cdot C_{sr}) + M_{SGSN} \cdot (U_{SGSN}^{(c)} + C_{rr} + 3 \cdot C_{gs})].
\end{aligned}$$

The total location update cost per unit time is the sum of the above two expressions

$$U_{II} = U_{II}^{(c)} + U_{II}^{(i)}.$$

Bibliography

3GPP (2002) Architectural requirements for Release 1999. TS 23.121, 3rd Generation Parnership Project (3GPP).

3GPP (2003) Multimedia Broadcast/Multicast Service (MBMS); Stage 2. TR 23.846, 3rd Generation Parnership Project (3GPP).

3GPP (2007a) 3GPP system to Wireless Local Area Network (WLAN) interworking; System description. TS 23.234, 3rd Generation Parnership Project (3GPP).

3GPP (2007b) General Packet Radio Service (GPRS); Service description; Stage 2. TS 23.060, 3rd Generation Parnership Project (3GPP).

3GPP (2007c) Generic Authentication Architecture (GAA); Generic bootstrapping architecture. TS 33.220, 3rd Generation Parnership Project (3GPP).

3GPP (2007d) Multimedia Broadcast/Multicast Service (MBMS); Architecture and functional description. TS 23.246, 3rd Generation Parnership Project (3GPP).

3GPP (2007e) Multimedia Broadcast/Multicast Service (MBMS); Protocols and codecs. TS 26.346, 3rd Generation Parnership Project (3GPP).

3GPP (2007f) Multimedia Broadcast/Multicast Service (MBMS) user services; Stage 1. TS 22.246, 3rd Generation Parnership Project (3GPP).

3GPP (2007g) Physical layer – general description. TS 25.201, 3rd Generation Parnership Project (3GPP).

3GPP (2007h) Quality of Service (QoS) concept and architecture. TS 23.107, 3rd Generation Parnership Project (3GPP).

3GPP (2007i) Transparent end-to-end Packet-Switched Streaming Service (PSS); Protocols and codecs. TS 26.234, 3rd Generation Parnership Project (3GPP).

3GPP (2008a) Introduction of the Multimedia Broadcast/Multicast Service (MBMS) in the Radio Access Network (RAN); Stage 2. TS 25.346, 3rd Generation Parnership Project (3GPP).

3GPP (2008b) Medium Access Control (MAC) protocol specification. TS 25.321, 3rd Generation Parnership Project (3GPP).

3GPP (2008c) Radio interface protocol architecture. TS 25.301, 3rd Generation Parnership Project (3GPP).

3GPP (2008d) Radio Link Control (RLC) protocol specification. TS 25.322, 3rd Generation Parnership Project (3GPP).

3GPP2 (2001) Wireless IP network standard. Technical Report P.S0001-A, 3rd Generation Parnership Project 2 (3GPP2).

3GPP2 (2002a) Interoperability specification (IOS) for CDMA2000 access network interfaces – Part 3; Features. Technical Report A.S0013-A, 3rd Generation Parnership Project 2 (3GPP2).

Multicast in Third-Generation Mobile Networks   Robert Rümmler, Alexander Gluhak and A. Hamid Aghvami
© 2009 John Wiley & Sons, Ltd

3GPP2 (2002b) Introduction to CDMA2000 Standards for Spread Spectrum Systems. Technical Report C.S0001-C, 3rd Generation Parnership Project 2 (3GPP2).

3GPP2 (2002c) IP Network Architecture Model for CDMA2000 Spread Spectrum Systems. Technical Report S.R0037, 3rd Generation Parnership Project 2 (3GPP2).

3GPP2 (2003) Broadcast Multicast Service Security Framework. Technical Report S.R0083-0, 3rd Generation Parnership Project 2 (3GPP2).

3GPP2 (2006a) CDMA2000 High Rate Broadcast-Multicast Packet Data Air Interface Specification. Technical Report C.S0054-A, 3rd Generation Parnership Project 2 (3GPP2).

3GPP2 (2006b) Interoperability Specification (IOS) for Broadcast Multicast Services (BCMCS). Technical Report A.S0019-A, 3rd Generation Parnership Project 2 (3GPP2).

3GPP2 (2007a) Broadcast and Multicast Service in CDMA2000 Wireless IP Network; Revision A. Technical Report X.S0022-A, 3rd Generation Parnership Project 2 (3GPP2).

3GPP2 (2007b) Interoperability Specification (IOS) for CDMA2000 Access Network Interfaces – Part 1; Overview. Technical Report A.S0011-D, 3rd Generation Parnership Project 2 (3GPP2).

3GPP2 (2008) Enhanced Cryptographic Algorithms. Technical Report S.S0055-A, 3rd Generation Parnership Project 2 (3GPP2).

Acharya S, Franklin M and Zdonik S (1995) Dissemination-based data delivery using broadcast disks. *IEEE Personal Communications* **2**(6), 50–60.

Adams A, Nicholas J and Siadak W (2005) Protocol Independent Multicast – Dense Mode (PIM-DM): protocol specification (revised). RFC 3973, Internet Engineering Task Force.

Agashe P, Rezaiifar R and Bender P (2004) CDMA2000 high rate broadcast packet data air interface design. *IEEE Communications Magazine* **42**(2), 83–89.

Aho K, Ristaniemi T, Kurjenniemi J and Haikola V (2007) System level performance of Multimedia Broadcast Multicast Service (MBMS) with macro diversity. *Wireless Communications and Networking Conference (WCNC 2007)*, Hong Kong, IEEE.

Alahuhta P, Jurvansuu M and Pentikäinen H (2004) Roadmap for network technologies and services. Technical report, Technical Research Centre of Finland VTT. Technology review 162/2004.

Alavi H and Nettleton R (1982) Downstream power control for a spread spectrum cellular mobile radio system. *Proceedings of GLOBECOM'82, Miami, Florida*.

Albanna Z, Almeroth K, Meyer D and Schipper M (2001) IANA guidelines for IPv4 multicast address assignments. RFC 3171, Internet Engineering Task Force.

Alexandri E, Amram N, Mueller R, van der Kreeft P and Lebeugle F (2006) Scenarios and technical requirements. Deliverable D2.1, C-Mobile.

Almeroth K (2000) The evolution of multicast: from the MBone to interdomain multicast to Internet2 deployment. *IEEE Network* **14**(1), 10–20.

Ballardie A (1997) Core Based Trees (CBT version 2) multicast routing – protocol specification. RFC 2189, Internet Engineering Task Force.

Bettstetter C, Vögel HJ and Eberspächer J (1999) GSM phase 2+ general packet radio service GPRS: architecture, protocols and air interface. *IEEE Communication Surveys* **2**(3), 2–14.

Bhattacharyya S (2003) An overview of Source-Specific Multicast (SSM). RFC 3569, Internet Engineering Task Force.

Bhushan N, Lott C, Black P, Attar R, Jou YC, Fan M, Ghosh D and Au J (2006) CDMA2000 1xEV-DO Revision A: a physical layer and MAC layer overview. *IEEE Communications Magazine* **44**(2), 75–87.

Blust SM (2002) SDR Forum roles and global work focus on radio software download. *IEICE Transactions on Communications* **E85-B**(12), 2581–2587.

Boehm B (2003) Value-based software engineering. *SIGSOFT Software Engineering Notes* **28**(2), 1–12.

Boivie R, Feldman N, Imai Y, Livens W and Ooms D (2007) Explicit Multicast (Xcast) concepts and options. RFC 5058, Internet Engineering Task Force.

Brand A and Aghvami H (2002) *Multiple Access Protocols for Mobile Communications: GPRS, UMTS and Beyond*. John Wiley.

Cain B, Deering S, Kouvelas I, Fenner B and Thyagarajan A (2002) Internet Group Management Protocol, Version 3. RFC 3376, Internet Engineering Task Force.

Campbell AT, Gomez J, Kim S, Wan CY, Turanyi ZR and Valko AG (2002) Comparison of IP micromobility protocols. *IEEE Wireless Communications* 9(1), 72–82.

Cardieri P and Rappaport TS (2001) Statistical analysis of co-channel interference in wireless communications systems. *Wireless Communications and Mobile Computing* 1, 111–121.

Chang J and Yang T (1994) End-to-end delay of an adaptive selective repeat ARQ protocol. *IEEE Transactions on Communications* 42, 2926–2928.

Chari M, Ling F, Mantravadi A, Krishnamoorthi R, Vijayan R, Walker GK and Chandhok R (2007) FLO physical layer: an overview. *IEEE Transactions on Broadcasting* 53(1), 145–160.

Choi W and Kim JY (2001a) Forward-link capacity of a DS/CDMA system with mixed multirate sources. *IEEE Transactions on Vehicular Technology* 50(3), 737–749.

Choi W and Kim JY (2001b) Joint Erlang capacity of DS/CDMA forward link based on resource sharing algorithm. *IEICE Transactions on Fundamentals* E84–A, 1406–1412.

Davis FD (1989) Perceived usefulness, perceived ease of use, and user acceptance of information technology. *MIS Quarterly* 13, 3.

Deering S (1989) Host extensions for IP multicasting. RFC 1112, Internet Engineering Task Force.

Deering S (1991) Multicast routing in a datagram internetwork. *PhD thesis*, Stanford, Palo Alto, CA.

Deering S, Estrin DL, Farinacci D, Jacobson V, Liu CG and Wei L (1996) The PIM architecture for wide-area multicast routing. *IEEE/ACM Transactions on Networking* 4(2), 153–162.

Deering SE and Cheriton DR (1990) Multicast routing in datagram internetworks and extended LANs. *ACM Transactions on Computer Systems* 8(2), 85–110.

Diot C, Levine BN, Lyles B, Kassem H and Balensiefen D (2000) Deployment issues for the IP multicast service and architecture. *IEEE Network* 14, 78–88.

DVB Project (2008) wwww.dvb.org.

Estrin D, Farinacci D, Helmy A, Thaler D, Deering S, Handley M, Jacobson V, Liu C, Sharma P and Wei L (1998) Protocol Independent Multicast-Sparse Mode (PIM-SM): protocol specification. RFC 2362, Internet Engineering Task Force.

ETSI (2001) Digital Audio Broadcasting (DAB); DAB to mobile, portable and fixed receivers. Technical report EN 300 401, European Telecommunications Standards Institute.

ETSI (2003) Digital Video Broadcasting (DVB); DVB specification for data broadcasting. Technical report EN 301 192, European Telecommunications Standards Institute.

ETSI (2004) Digital Video Broadcasting (DVB); Transmission system for handheld terminals (DVB-H). Technical report EN 302 304, European Telecommunications Standards Institute.

ETSI (2005) Digital Audio Broadcasting (DAB); DMB video service; User application specification. Technical report TS 102 428, European Telecommunications Standards Institute.

ETSI (2007) Digital Video Broadcasting: IP datacast over DVB-H: Set of specifications for phase 1. Technical report TS 102 428, European Telecommunications Standards Institute.

Evans JS and Everitt D (1999) On the teletraffic capacity of CDMA cellular networks. *IEEE Transactions on Vehicular Technology* 48(1), 153–165.

Faria G, Henriksson JA, Stare E and Talmola P (2006) DVB-H digital broadcast services to handheld devices. *Proceedings of the IEEE* 94(1), 194–209.

Fenner B and Meyer D (2003) Multicast Source Discovery Protocol (MSDP). RFC 3618, Internet Engineering Task Force.

Fenner W (1997) Internet Group Management Protocol, Version 2. RFC 2236, Internet Engineering Task Force.

Floyd S, Jacobson V, Liu CG, McCanne S and Zhang L (1997) A reliable multicast framework for light-weight sessions and application level framing. *IEEE/ACM Transactions on Networking* **5**(6), 784–803.

Fogg BJ and Tseng H (1999) The elements of computer credibility. *Proceedings of CHI99 Conference*, New York, USA pp. 80–87.

Gefen D, Karahanna E and Straub DW (2003) Inexperience and experience with online stores: the importance of TAM and Trust. *IEEE Transactions on Engineering Management* **50**(3), 307–321.

Gejji R (1992) Forward-link power control in cellular CDMA systems. *IEEE Transactions on Vehicular Technology* **41**, 1195–1197.

Gilhousen KS, Jacobs IM, Padovani R, Viterbi AJ, Weaver LA and Wheatley III CE (1991) On the capacity of a cellular CDMA system. *IEEE Transactions on Vehicular Technology* **40**(2), 303–312.

Goodman D (1990) Cellular packet communications. *IEEE Transactions on Communications* **38**(8), 1272–1280.

Group AVTW, Schulzrinne H, Casner S, Frederick R and Jacobson V (1996) RTP: a transport protocol for real-time applications. RFC 1889, Internet Engineering Task Force.

Haberman B and Thaler D (2002) Unicast-prefix-based IPv6 multicast addresses. RFC 3306, Internet Engineering Task Force.

Handley M and Jacobson V (1998) SDP: Session Description Protocol. RFC 2327, Internet Engineering Task Force.

Harrison T, Williamson C, Mackrell W and Bunt R (1997) Mobile Multicast (MoM) Protocol: multicast support for mobile hosts. *ACM MobiCom 1997*.

Hinden R and Deering S (1998) IPv6 multicast address assignments. RFC 2375, Internet Engineering Task Force.

Ho JS and Akyildiz IF (1996) Local anchor scheme for reducing signaling costs in personal communications networks. *IEEE/ACM Transactions on Networking* **4**(5), 709–725.

Holbrook H and Cain B (2006) Source-specific multicast for IP. RFC 4607, Internet Engineering Task Force.

Holma H and Toskala A (eds) (2007) *WCDMA for UMTS – HSPA Evolution and LTE*. John Wiley & Sons.

Ishikawa Y, Hayashi T and Onoe S (2002) W-CDMA downlink transmit power and cell coverage planning. *IEICI Transactions on Communications* **E85-B**(11), 2416–2426.

Jakes WC (ed.) (1994) *Microwave Mobile Radio Communications*. IEEE Press, New York, NY.

Jansen M and Prasad R (1995) Capacity, throughput, and delay analysis of a cellular DS CDMA system with imperfect power control and imperfect sectorization. *IEEE Transactions on Vehicular Technology* **44**(1), 67–75.

Jelger C and Noel T (2002) Multicast for mobile hosts in IP networks: progress and challenges. *IEEE Wireless Communications* **9**(5), 58–64.

Kaarannen H, Ahtiainen A, Laitinen L, Naghian S and Niemi V (eds) (2001) *UMTS Networks: Architecture, Mobility and Services*. John Wiley & Sons.

Kaasinen E (2005) *User acceptance of mobile services – value*, ease of use, trust and ease of adoption. *PhD thesis*, Tampere University of Technology, Tampere, FL.

Karn P, Bormann C, Fairhurst G, Grossman D, Ludwig R, Mahdavi J, Montenegro G, Touch J and Wood L (2004) Advice for Internet subnetwork designers. RFC 3819, Internet Engineering Task Force.

Kim J and Krunz M (2000) Delay analysis of selective repeat ARQ for a Markovian source over a wireless channel. *IEEE Transactions on Vehicular Technology* **49**, 1968–1981.

Kim SW, Jeong DG, Jeon WS and Choi CH (2002) Forward link performance of combined soft and hard handoff in multimedia CDMA systems. *IEICE Transactions on Communications* **E85-B**(7), 1276–1281.

Koodli R and Puuskari M (2001) Supporting packet-data QoS in next-generation cellular networks. *IEEE Communications Magazine* **39**(2), 180–188.

Kornfeld M and May G (2007) DVB-H and IP datacast – broadcast to handheld devices. *IEEE Transactions on Broadcasting* **53**(1), 161–170.

Kosiur D (1998) *IP Multicast: The Complete Guide to Interative Corporate Networks*. John Wiley & Sons.

Lee CC and Steele R (1998) Effect of soft and softer handoffs on CDMA system capacity. *IEEE Transactions on Vehicular Technology* **47**(3), 830–841.

Lee J (2001) Multicast Avalanche Avoidance in Mobile IP (MAAMIP). Internet-Draft draft-lee-maa-mip-00.txt, Internet Engineering Task Force.

Lee JS and Miller LE (1999) Solutions for minimum required forward link channel powers in CDMA cellular and PCS systems. *Journal of Communications and Networks* **1**(1), 42–51.

Lee W (1991a) Power control in CDMA. *Proceedings of IEEE VTC 1991*, St. Louis, Missouri. pp. 77–80.

Lee WCY (1991b) Overview of cellular CDMA. *IEEE Transactions on Vehicular Communications* **40**(2), 291–302.

Li V and Zhang Z (2002) Internet multicast routing and transport control protocols. *Proceedings of the IEEE* **90**(3), 360–391.

Lin C and Wang K (2000) Mobile multicast support in IP networks. *IEEE INFOCOM 2000*, Tel Aviv, Israel, IEEE.

Lin JC and Paul S (1996) RMTP: a reliable multicast transport protocol. *IEEE INFOCOM '96*, 15th Annual Joint Conference of the IEEE Computer Society, San Francisco, CA, Vol. **3**, pp. 1414–1424.

Lin YB (2001) A multicast mechanism for mobile networks. *IEEE Communications Letters* **5**(11), 450–452.

Lin YB and Chen YK (2003) Reducing authentication signaling traffic in third generation mobile network. *IEEE Transactions on Wireless Communications* **2**(3), 493–501.

Lin YB, Haung YR, Chen YK and Chlamtac I (2001) Mobility management: from GPRS to UMTS. *Wireless Communications and Mobile Computing* **1**(4), 339–359.

Lin YB, Lee PC and Chlamtac I (2002) Dynamic periodic location area update in mobile networks. *IEEE Transactions on Vehicular Technology* **51**(6), 1494–1501.

Lin YB, Rao HCH and Chlamtac I (2001) General packet radio service (GPRS): architecture, interfaces, and deployment. *Wireless Communications and Mobile Computing* **1**(1), 77–92.

Linden A and Fenn J (2003) Understanding Gartner's hype cycles. Technical report R-20-1971, Gartner Research.

Luby M, Gemmell J, Vicisano L, Rizzo L and Crowcroft J (2002a) Asynchronous Layered Coding (ALC) Protocol instantiation. RFC 3450, Internet Engineering Task Force.

Luby M, Gemmell J, Vicisano L, Rizzo L, Handley M and Crowcroft J (2002b) Layered Coding Transport (LCT) building block. RFC 3451, Internet Engineering Task Force.

Luby M, Vicisano L, Gemmell J, Rizzo L, Handley M and Crowcroft J (2002c) The use of Forward Error Correction (FEC) in reliable multicast. RFC 3453, Internet Engineering Task Force.

McAuley AJ (1990) Reliable broadband communications using a burst erasure correcting code *Proceedings of ACM SIGCOMM'90*, Philadelphia, PA, pp. 287–306.

McCann PJ and Hiller T (2000) An Internet infrastructure for cellular CDMA networks using Mobile IP. *IEEE Personal Communications* **7**(4), 6–12.

Miller K, Robertson K, Tweedly A and White M (1997) StarBurst Multicast File Transfer Protocol (MFTP) specification. Internet-Draft draft-miller-mftp-spec-02.txt, Internet Engineering Task Force.

Mohan S and Jain R (1994) Two user location strategies for personal communication services. *IEEE Personal Communications* **1**(1), 42–50.

MorganDoyle (2001) The mobile data value chain. Technical report, MorganDoyle Limited.

Moy J (1994) Multicast extensions to OSPF. RFC 1584, Internet Engineering Task Force.

Murphy T (2001) The CDMA2000 packet core network. *Ericsson Review* (2), 88–95.

Nonnenmacher J (1998) Reliable multicast over large groups. *PhD thesis*, Section of Communication Systems, Swiss Federal Institute of Technology (EPFL), Lausanne, Switzerland.

OMA (2007) Push architecture. Technical Report OMA-AD-Push-V2_2-20071002-C, Open Mobile Alliance.

Paila T, Luby M, Lehtonen R, Roca V and Walsh R (2004) FLUTE – file delivery over unidirectional transport. RFC 3926, Internet Engineering Task Force.

Perkins C (1996) IP mobility support. RFC 2002, Internet Engineering Task Force.

Perkins C (2002) IP mobility support for IPv4. RFC 3220, Internet Engineering Task Force.

Picard RG (1998) Interacting forces in the development of communication technologies. *European Media Management Review* **1**(1), 18–24.

Picard RG (2005) Mobile telephony and broadcasting: are they compatible for consumers. *Int. J. Mobile Communications* **3**(1), 19–28.

Plummer D (1982) Ethernet Address Resolution Protocol: or converting network protocol addresses to 48.bit Ethernet addresses for transmission on Ethernet hardware. RFC 0826, Internet Engineering Task Force.

Poole I (2004/2005) What exactly is Mobile IP? *IEE Communications Engineer* **12**(4), 44–45.

Qualcomm (2007) FLO technology overview. Technical report, Qualcomm Incorporated.

Rizzo L (1997) Effective erasure codes for reliable computer communication protocols. *Computer Communication Review* **27**(2), 24–36.

Rosenberg J, Schulzrinne H, Camarillo G, Johnston A, Peterson J, Sparks R, Handley M and Schooler E (2002) SIP: Session Initiation Protocol. RFC 3261, Internet Engineering Task Force.

Rossi M, Badia L and Zorzi M (2003a) Exact statistics of ARQ packet delivery delay over Markov channels with finite round-trip delay. *Proceedings of IEEE Globecom 2004*, San Francisco, CA, pp. 3356–3360.

Rossi M, Badia L and Zorzi M (2003b) On the delay statistics of an aggregate of SR-ARQ packets over Markov channels with finite round-trip delay. *IEEE WCNC*, New Orleans, LA.

Rossi M, Badia L and Zorzi M (2005) On the delay statistics of SR ARQ over Markov channels with finite round-trip delay. *IEEE Transactions on Wireless Communications* **4**(4), 1858–1868.

Rossi M, Fitzek F and Zorzi M (2003) Error control techniques for efficient multicast streaming in UMTS networks. *SCI Conference*, Orlando, FL.

Rossi M, Zorzi M and Fitzek F (2004) Link layer algorithms for efficient multicast service provisioning in 3G cellular systems. *IEEE Globecom*, Dallas, TX.

Royce W (1970) Managing the development of large software systems. *Proceedings of IEEE WESCON, Los Alamitos, California, 1970*.

Sahasrabuddhe LH and Mukherjee B (2000) Multicast routing algorithms and protocols: a tutorial. *IEEE Network* **14**(1), 90–102.

Salkintzis AK (2001) Wireless IP with GPRS: fundamental operational aspects. *The Fourth International Symposium on Wireless Personal Multimedia Communications*, Aalborg, Denmark, pp. 7–15.

Schulzrinne H, Rao A and Lanphier R (1998) Real Time Streaming Protocol (RTSP). RFC 2326, Internet Engineering Task Force.

Shokrollahi A (2006) Raptor codes. *IEEE Transactions on Information Theory* **52**(6), 2551–2567.

Simpson F and Holtzman J (1993) Direct sequence CDMA power control, interleaving, and coding. *IEEE Journal on Selected Areas in Communications* **11**(7), 1085–1095.

Slimane SB (2001) Bounds on the distribution of a sum of independent lognormal random variables. *IEEE Transactions on Communications* **49**(6), 975–978.

Speakman T, Crowcroft J, Gemmell J, Farinacci D, Lin S, Leshchiner D, Luby M, Montgomery T, Rizzo L, Tweedly A, Bhaskar N, Edmonstone R, Sumanasekera R and Vicisano L (2001) PGM Reliable Transport Protocol specification. RFC 3208, Internet Engineering Task Force.

Steele R and Hanzo L (eds) (1999) *Mobile Radio Communications, Second and Third Generation Cellular and WATM Systems*, 2nd edition. John Wiley & Sons, Ltd, Chichester, UK.

Stevens W (1994) *TCP/IP Illustrated, Vol. 1: The Protocols*. Addison Wesley Longman.

Suh Y, Shin H and Hee D (2001) An efficient multicast routing protocol in wireless mobile networks. *Wireless Networks* **7**, 443–453.

Takada M and Saito M (2006) Transmission system for ISDB-T. *Proceedings of the IEEE* **94**(1), 251–256.

Tam WM and Lau FCM (1999) Analysis of power control and its imperfections in CDMA cellular systems. *IEEE Transactions on Vehicular Technology* **48**(5), 1706–1717.

Thaler D (2004) Border Gateway Multicast Protocol (BGMP): protocol specification. RFC 3913, Internet Engineering Task Force.

Thaler D, Handley M and Estrin D (2000) The Internet multicast address allocation architecture. RFC 2908, Internet Engineering Task Force.

The Global Mobile Suppliers Association (2008) GSM/3G network update. Technical report, The Global mobile Suppliers Association.

TIA (2006) Forward link only air interface specification for terrestrial mobile multimedia multicast. Technical report 1099, Telecommunications Industry Association.

TIA (2007) Forward link only transport specification. Technical report 1120, Telecommunications Industry Association.

Tuttlebee W, Babb D, Irvine J, Martinez G and Worrall K (2003) Broadcasting and mobile telecommunications: interworking – not convergence. *European Broadcasting Union Technical Review* (293).

Tuttlebee W and Payne M (2004/2005) The giant answers the call. *IEE Communications Engineer* **2**(6), 28–31.

Ulukus S and Yates RD (1998) Stochastic power control for cellular radio systems. *IEEE Transactions on Communications* **46**(6), 784–798.

Viterbi A, Gilhousen K and Zehavi E (1994) Soft handoff extends CDMA cell coverage and increases reverse link capacity. *IEEE Journal on Selected Areas in Communications* **12**(8), 1281–1288.

Viterbi AJ (1995) *CDMA Principles of Spread Spectrum Communication* **2**(6) Addison-Wesley.

Waitzman D, Partridge C and Deering S (1988) Distance Vector Multicast Routing Protocol. RFC 1075, Internet Engineering Task Force.

Watson M, Luby M and Vicisano L (2007) Forward Error Correction (FEC) building block. RFC 5052, Internet Engineering Task Force.

Whetten B, Vicisano L, Kermode R, Handley M, Floyd S and Luby M (2001) Reliable multicast transport building blocks for one-to-many bulk-data transfer. RFC 3048, Internet Engineering Task Force.

Wikipedia (2008a) 3G, http://en.wikipedia.org/wiki/3G.

Wikipedia (2008b) Cell broadcast, http://en.wikipedia.org/wiki/Cell_Broadcast.

Williams A (2004) The documentation of quality engineering: applying use cases to driver change in software engineering models *Proceedings of the 22nd Annual International Conference on Design of Communication: the Engineering of Quality Documentation*.

Wong VWS and Leung VCM (2000) Location management for next-generation personal communication networks. *IEEE Network* **14**(5), 18–24.

Xie J and Akyildiz IF (2002) A novel distributed dynamic location management scheme for minimizing signaling costs in Mobile IP. *IEEE Transactions on Mobile Computing* **1**(3), 163–175.

Zander J (1993) Transmitter power control for co-channel interference management in cellular radio systems. *Proceedings of 4th WINLAB Workshop*, New Brunswick, NF.

Zanoio E and Urvik S (2003) CDMA network technologies: a decade of advances and challenges. Technical report, Tektronix.

Zhang X, Castellanos JG and Campbell AT (2002) P-MIP: Paging extensions for Mobile IP. *Mobile Networks and Applications* **7**(2), 127–141.

Zimmermann H (1980) OSI reference model – The OSI model of architecture for open systems interconnection. *IEEE Transactions on Communications* **COM-28**(4), 425–432.

Zorzi M (1995) Fast computation of outage probability for cellular mobile radio systems. *European Transactions on Telecommunications* **6**(1), 107–113.

Zorzi M (1996) Power control and diversity in mobile radio cellular systems in the presence of Ricean fading and log-normal shadowing. *IEEE Transactions on Vehicular Technology* **45**(2), 373–382.

Zorzi M and Milstein LB (1994) Power control and coding in a cellular CDMA mobile radio system. *Proceedings of IEEE ICC'94*, New Orleans, Louisiana.

Zorzi M, Rossi M and Mazzini G (2002) Throughput and energy performance of TCP on a wideband CDMA air interface. *Wireless Communications and Mobile Computing* **2**(1), 71–84.

Index

Multicast in Third-Generation Mobile Networks Robert Rümmler, Alexander Gluhak and A. Hamid Aghvami
© 2009 John Wiley & Sons, Ltd